Rock Fall
Engineering

Rock Fall Engineering

DUNCAN C. WYLLIE

CRC Press
Taylor & Francis Group
Boca Raton London New York

CRC Press is an imprint of the
Taylor & Francis Group, an **informa** business

CRC Press
Taylor & Francis Group
6000 Broken Sound Parkway NW, Suite 300
Boca Raton, FL 33487-2742

First issued in paperback 2017

© 2015 by Taylor & Francis Group, LLC
CRC Press is an imprint of Taylor & Francis Group, an Informa business

No claim to original U.S. Government works
Version Date: 20140814

ISBN 13: 978-1-138-07762-1 (pbk)
ISBN 13: 978-1-4822-1997-5 (hbk)

Visit the Taylor & Francis Web site at
http://www.taylorandfrancis.com

and the CRC Press Web site at
http://www.crcpress.com

Contents

About the Author

Duncan C. Wyllie earned a physics degree from the University of London, and engineering degrees from the University the New South Wales, Australia, the University of California, Berkeley, and the University of British Columbia. He is presently a principal with Wyllie & Norrish Rock Engineers in Vancouver, Canada, a specialist engineering company working in the fields of rock slopes, landslides, tunnels, foundations, and blasting, and is a registered Professional Engineer in British Columbia, Canada.

Duncan has been working in the field of applied rock mechanics since the mid-1960s, for both civil and mining projects. These projects have been undertaken mainly in North America but also in overseas countries from Australia to Turkey. Typical assignments have included the design and construction of slopes and tunnels, foundations of bridges and dams, and the study of landslides. He carried out the initial development of a widely used hazard rating system for highways in mountainous terrain. He has also worked in the mining industry in the design of open-pit slopes, underground support, and tailings dams.

Duncan has lectured widely for 30 years, conducting training courses in rock slope engineering to state and federal highway engineers across the United States and fourth-year courses in engineering geology at the University of British Columbia. He has also authored and coauthored a number of textbooks on applied rock mechanics: *Foundations on Rock* (1st and 2nd editions in 1989 and 2001), *Rock Slope Engineering* (4th edition, 2002), both also published by Taylor & Francis, and *Landslides, Investigation and Mitigation* published by the Transportation Research Board in 2004.

Since 2009, he has been conducting research in rock falls, with particular emphasis on the application of impact mechanics to rock fall behavior, and how this can be used to model rock falls and design protection structures. The results of this research, together with nearly 40 years of experience on projects involving rock falls, are the subject of this book.

Introduction

This book on rock fall engineering has arisen from an initial passing interest in the subject as the result of extensive project work on transportation projects in the mountainous area of North America. This interest developed into a mission to fully understand all aspects of rock fall behavior and the application of this behavior to the design and construction of protection structures. Lately, this mission has evolved into an obsession to help develop improved methods of modeling rock falls and the design of more efficient and cost-effective protection structures.

As with my other two books, *Foundations on Rock* and *Rock Slope Engineering*, the intention of this book is to provide both the theory, and the application of the theory to design. In this book, this approach involves describing five case studies where the impacts are well defined, and then showing how trajectory calculations and impact mechanics can be applied to these actual rock fall conditions. It is hoped that the field data will be useful for calibration of computer rock fall simulation programs.

In addition, a wide range of well-proven rock fall protection measures are discussed. These discussions describe both design methods, and practical construction experience based on many projects in which the author has been involved. It is intended that users of this book will be both researchers working on the development of rock fall simulation, and practitioners working in the field of rock fall mitigation design and construction.

My work on rock falls has benefited from my association over many years with practitioners involved with the design and construction of mitigation structures. These people include Dale Harrison, Chuck Brawner, and British Columbia Ministry of Transport personnel in Canada; John Duffy in California; Bob Barrett, Rick Andrew, Randy Jibson (USGS); and Ty Ortiz (CDOT) in Colorado. In Japan, I have worked closely with Toshimitsu Nomura and his colleagues at Protec Engineering, as well as Dr. Masuya at Kanazawa University and Dr. Hiroshi Yoshida. Dr. Bill Stronge of Cambridge University has also been most helpful in furthering my understanding of impact mechanics as well as Dr. Giacomini at Newcastle University in Australia.

I also acknowledge my long association with the Canadian Pacific Railway and their rock mechanics engineer Tony Morris. Many of the concepts discussed in this book have developed from a wide variety of rock fall protection projects that the railway has undertaken.

The research on rock falls and the preparation of this book have involved the collection of field data and analysis of the results. I have received much valuable assistance in this work from Thierry Lavoie, Phillip Lesueur, and Tom Beingessner while they were attending the University of British Columbia. In addition, Tom Reynolds conducted the model tests of attenuator nets and canopies, Jan Meyers has spent long hours compiling references and research documents, and Rhona Karbusicky found many vital and sometimes obscure references. Most important, Cheng-Wen Tina Chen has provided invaluable assistance in data analysis and preparing drawings and in overall organization and preparation of the manuscript.

Some early funding for this work was provided by the National Research Council of Canada, and the Railway Ground Hazards Research Group in Alberta, Canada, for which I am appreciative.

I would also like to acknowledge the support of my family during yet another long period of dedication to book writing.

Duncan C. Wyllie
North Vancouver, Canada

Foreword

After having published two highly appreciated books on geological engineering for rock slopes, then for rock foundations, Duncan Wyllie now presents a fine instructive volume on geological engineering for protecting rock cuts and engineering works potentially threatened by rock falls. This richly illustrated treatise improves the practical tools available for evaluating and remediating potential rock fall hazards that can threaten highways and developments with high-velocity rock mass impacts. This work is not intended to be a survey of various geologies and methodologies, but rather functions as a practical tutor for geologists and engineers evaluating rock fall hazards and engineering safeguards.

The book is clearly written and concise, and contains ideas refining and focusing the analytical treatment of rock fall paths, and energies, including an introduction to the fundamentals of impact mechanics. There follows a discussion of the kinematics and energy balances of bouncing trajectories affecting a rock fall's path and velocity as well as a detailed discussion of coefficients of restitution. The attention given to these important facets of rock fall engineering, which may be new to some readers, gives special value to this book.

The importance of accurately accounting for energy gains and losses during rock fall descent is very much in my mind as I recall an experience observing a series of rock fall tests at an abandoned quarry being considered for home construction. Large blocks of rock were trucked to the quarry top and released. Each block path was filmed up to its final landing. Observing from below at "a very safe distance," I witnessed some surprising block trajectories in which a large block bouncing from a rock shelf would explosively release a host of smaller block fragments on entirely new paths. After I left the site, the experiment continued and several large block pieces over-flew my previous "safe" station.

The issue of rock fall protection is well presented, with chapters on selection and design of rock fall protection ditches, barriers, nets, fences, and rock sheds. Very importantly, the application of popular computational systems for rock fall modeling is evaluated in the light of five instructive case histories.

Richard E. Goodman
Emeritus Professor of Geological Engineering
University of California, Berkeley, CA

Nomenclature

0 Subscript for velocities at the start of trajectory ($t = 0$)
A Width of MSE barrier at impact height (m)
A Constant used in [time–force] relationship for flexible nets; acceleration
B Base width of MSE barrier (m)
B Constant used in [time–force] relationship for stiff nets
C Coefficient related to mode of failure of rock sheds; crest width of MSE barrier (m)
D Diameter of falling rock (m)
d_b Drill hole diameter (mm)
e_N Normal coefficient of restitution
e_T Tangential coefficient of restitution
E_c Energy absorbed during compression phase of impact (J)
E_e Energy efficiency for fence design
$(E_f–E_c)$ Energy recovered during restitution phase of impact (J)
E_i, E_f Impact (i) and restitution (final, f) energies for impact with protection structures (J)
F Force (N)
f Subscript for velocities and energies at the completion of impact ($t = f$)
g Gravitational acceleration (m · s^{-2})
H Rock fall height (m)
h Trajectory height—vertical (m)
h' Trajectory height—normal to slope (m)
I Moment of inertia (kg · m^2)
I' Tensor defining components of moments of inertia
i Subscript for velocities at the moment of impact ($t = i$); inclination of asperities (degrees)
k Radius of gyration (m)
L Side length of cubic block; length of trajectory between impacts; bond length of rock bolts; sliding length of rock falls (m)
M Average mass of rock falls related to Gumbel extreme value theory
m Mass of rock fall (kg)
$m_{(n)}$ Mass of rock fall at impact point n (kg)
$m_{(0)}$ Mass of rock fall at source (kg)
N Subscript for the component of velocity normal to the slope
n Impact number; gradient of line for [time–force] relationship for rigid structures
P Equivalent static force in roof of rock sheds (kN)
p Probability
p_N Normal impulse (kg · m · s^{-1})
p_T Tangential impulse (kg · m · s^{-1})
R Frictional resistance at impact point
r Radius of rock fall body (m)

S Sliding distance (m); standard deviation of mass of rock falls related to Gumbel extreme value theory

s Dimension defining slope roughness (m)

T Subscript for the component of velocity tangential to the slope; thickness of sand cushion on rock fall sheds (m)

t Time (s)

v Relative velocity at contact point (m $\cdot$ s^{-1})

v_N Normal component of relative velocity at contact point (m $\cdot$ s^{-1})

v_T Tangential component of relative velocity at contact point (m $\cdot$ s^{-1})

V_i Velocity of centre of mass at impact time $t = 0$ (m $\cdot$ s^{-1})

V_f Velocity of centre of mass, final or restitution at time $t = f$ (m $\cdot$ s^{-1})

V_{iN} Normal component of impact velocity of centre of mass (m $\cdot$ s^{-1})

V_{iT} Tangential component of impact velocity of centre of mass (m $\cdot$ s^{-1})

V_{fN} Normal component of final velocity of centre of mass (m $\cdot$ s^{-1})

V_{fT} Tangential component of final velocity of centre of mass (m $\cdot$ s^{-1})

W Weight of sliding block (N)

x Horizontal coordinate (m); exponent in time–force power relationship

z Vertical coordinate (m)

α Angle of velocity vector relative to positive x-axis (degrees); location parameter (Gumbel extreme value distribution)

$\beta_1, \beta_2, \beta_3$ Inertial coefficients related to rotation of block during impact; scale parameter (Gumbel extreme value distribution); cushion layer thickness/rock fall diameter ratio

γ Factor of safety, fence design; density (kN $\cdot$ m^{-3})

δ Deformation or displacement or compression (mm)

δ_m, δ_v Displacement of mountain (m) and valley (v) sides of MSE banner

ε Angle defining slope roughness (m)

η Slope resistance factor used in velocity calculations

θ_i Impact angle relative to slope surface (degrees)

θ_f Final or restitution angle relative to slope surface (degrees)

K Slope gradient, trajectory calculations

λ Reduction coefficient related to loss of mass during rock falls (m^{-1}); Lamé parameter for sand cushion (kN $\cdot$ m^{-2})

μ Friction coefficient at impact point

μ' Effective friction coefficient of slope surface

$\sigma_{u(r)}$ Uniaxial compressive strength of rock (MPa)

τ_{all} Allowable rock-grout bond strength (kPa)

φ Friction angle (degrees)

ψ Dip angle–slope (s), face (f), plane (p), (degrees)

Ω Volume of rock fall (m^3)

Ω_0 Volume of rock fall at source (m^3)

ω Angular velocity (rad $\cdot$ s^{-1})

Chapter 1

Rock Falls—Causes and Consequences

In mountainous terrain, infrastructure such as highways, railways, and power generation facilities, as well as houses and apartment buildings, may be subject to rock fall hazards. These hazards can result in economic losses due to service interruptions, equipment damage, and loss of life. Rock fall hazards are particularly severe in areas with heavy precipitation, frequent freeze-thaw cycles, and seismic events (TRB, 1996). These climatic conditions exist, for example, in the Alps, on the West Coast of North America, and in Japan. In contrast, in Hong Kong, where temperatures are more mild but intense rainfall events occur, rock fall risks can also be severe because of the high population density (Chau et al., 2003).

Protection against rock fall hazards can be provided by a variety of structures that are now well proven as the result of extensive testing by the manufacturers of these systems and their use in a wide variety of conditions, as discussed in Chapters 10 and 11. These protection structures include ditches that can be designed to reasonably well-defined criteria, and will be more effective if they incorporate barriers with steep faces such as gabions or MSE walls constructed from locally available materials. In addition, proprietary fence systems have been developed that use various configurations of high-strength steel cables and wires. In some high-hazard locations, it may be appropriate to construct reinforced concrete rock sheds that incorporate energy-absorbing features such as flexible hinges and a cushioning layer of sand or Styrofoam on the roof (Japan Road Assoc., 2000; Yoshida et al., 2007).

Design of protection measures requires data for two basic parameters of rock falls—impact energy and trajectory. That is, information is required on the mass and velocity of falls to determine the required energy capacity, and on impact locations and trajectory paths to determine the optimum location and dimensions of the barrier or fence. Development of these design parameters requires the collection of relevant site data, followed by analysis of energies and trajectories and then selection and design of the appropriate protection measure.

The design process for protection structures comprises the following steps as described in this book:

- **Topography and geology**—The location of potential rock falls requires mapping to identify source areas, and the gullies in which the falls may concentrate. Geological studies will provide information on the likely size and shape of falls based on rock strength and on discontinuity persistence and spacing (Chapter 1).
- **Calibration of rock fall models**—Because of the complexity of rock fall behavior, it is useful to have data on actual rock falls with which to calibrate mathematical models. Falls on slopes comprising rock, talus, colluvium, asphalt, and concrete have been documented to provide this calibration data (Chapter 2).
- **Trajectory analysis**—The trajectory that the rock fall follows between impacts is a parabolic path defined by gravitational acceleration, resulting in translational energy

gain during the trajectory phase of the fall. Trajectories define the required height of protection structures (Chapter 3).

- **Impact mechanics**—The impact process between a rock fall and the slope can be defined by the theory of impact mechanics. Application of this theory to rock falls enables calculation of changes in tangential, normal, and rotational velocities that occur during impact, and the corresponding changes in kinetic and rotational energy (Chapter 4).
- **Coefficients of restitution**—The basic parameters defining the changes in tangential and normal velocities during impact are the corresponding coefficients of restitution. These parameters are related, respectively, to friction on the contact surface and the angle at which the body impacts the slope (Chapter 5).
- **Energy losses during impact**—The result of velocity changes during impact are corresponding changes to the translational and rotational energies of the body. The energy changes are the result of the frictional resistance to slipping/rolling in the tangential direction, and plastic deformation of the body and slope in the normal direction (Chapter 6).
- **Rock fall modeling**—Computer programs have been developed (by others) to model rock fall behavior and provide ranges of energies and trajectories for use in design. The principles of modeling are discussed, and the case studies described in Chapter 2 have been simulated in a widely used commercial rock fall modeling program to determine the parameters required to reproduce these actual events (Chapter 7).
- **Selection of protection structures**—Selection of the appropriate protection structure for a site involves first having a rational means, such as decision analysis, of selecting the required level of protection. Selection of rock mass values to use in design may involve statistical methods to extrapolate limited field data on rock fall dimensions. This analysis calculates the frequency of occurrence of design blocks with masses larger than those observed in the field (Chapter 8).
- **Design principles of protection structures**—Optimizing the absorption of impact energy by fences is related to attenuation in which the rock fall is deflected by the net, and the energy is absorbed uniformly over the time of impact. These attributes will limit impact forces generated in the structure (Chapter 9).
- **Protection structures**—Methods of protecting against rock falls include ditches, barriers, fence, nets, and rock sheds. Each structure has a specific range of impact energy capacity and suitability to the topography at the site, such as ditches, barriers, fences (Chapter 10), and rock sheds (Chapter 11).

1.1 SOURCE ZONES AND TOPOGRAPHY

Identification of rock fall source zones usually requires careful field investigations, possibly involving examination of air photographs, helicopter inspections, and climbing the slopes. Evidence of recent rock falls may include open tension cracks and fresh exposures on the rock faces, impact marks on trees along the fall path, and accumulations of falls on the lower part of the slope. It is also found that falls tend to collect in gullies, in the same way that water flows down valleys. That is, falls from a large area of the slope will accumulate at the base of gullies, a condition that can allow protection structures to be located only at these topographic features.

Other factors influencing rock fall behavior are the slope angle and the slope material. Figure 1.1 shows a typical slope configuration and the corresponding rock fall behavior on four zones of the slope as follows:

- **Rock slope**—On steep, irregular rock slopes, falls will have widely spaced impacts, high-speed translational and rotational velocities, and high-angle trajectories.
- **Colluvium slope**—On slopes that are just steeper than the angle of repose (i.e., if greater than 37 degrees for loose rock fragments), closely spaced impacts and shallow trajectories will occur, but falls will not accumulate on the slope.
- **Talus slope**—Falls accumulating on talus slopes form at the angle of repose ranging from about 37 degrees in the upper portion to 32 degrees near the base. Rock falls undergo a natural sorting when they reach the talus with smaller fragments accumulating near the top and larger ones reaching the base, such that the talus deposit enlarges uniformly forming a cone-shaped deposit.
- **Run-out zone**—A few of the larger, higher energy blocks may move beyond the base of the talus and on to a slope that is flatter than the talus. It has been found that the maximum run-out distance for these blocks is defined by a line inclined at about 27.5 degrees from the base of the steep rock slope; this angle represents the rolling friction coefficient of rock falls (Hungr and Evans, 1988). Within the run-out zone, rock falls move in a series of closely spaced impacts or rolling action, which means rocks can be readily stopped in this zone with shallow ditches or low fences.

The run-out zone as defined in Figure 1.1 has important implications for identifying hazards zones below rock slopes, and the need to install protection measures and/or establish development exclusion zones. Objects at risk that may be found within run-out zones include roads with low traffic volumes or golf courses that require little or no protection, to houses with full-time occupants that require high-reliability protection measures such as fences or barriers.

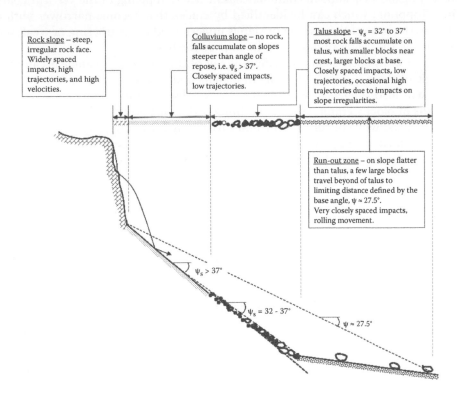

Figure 1.1 Typical slope configuration showing the relationship between slope angle and rock fall behavior.

1.2 GEOLOGY

Rock fall hazards are clearly related to the geology of the slope that is the potential source of falls. That is, the rock must be sufficiently strong to form a block that will survive impacts during the fall and not break into harmless fragments. For example, slopes in strong rocks such as granite, limestone, and basalt can be a source of dangerous rock falls, while weaker rocks such as shales and phyllites usually weather/fracture into small fragments that will not be a hazard. Another geological factor influencing rock fall hazards is discontinuity spacing and persistence. That is, closely spaced joints will form small blocks of rock with dimensions of a few centimetres, which may only be a hazard to particular vulnerable conditions such as pedestrian walkways and possibly automobiles. In contrast, more robust objects such as trains and power facilities may only be at risk from blocks with dimensions that exceed about 1 m (3.3 ft).

For slopes in strong rock, stability and rock fall occurrence are mainly related to the characteristics of the discontinuities. That is, joints dipping out of the face can form planar- or wedge-shaped blocks that may slide from the face, while joints dipping into the face can form toppling blocks (Wyllie and Mah, 2002).

Figure 1.2 shows typical geological conditions that can result in rock falls. For these three conditions, either surficial weathering or the orientation of discontinuities has exposed blocks of rock on the face that are vulnerable to instability.

Figure 1.2(a) shows a common condition in horizontally bedded sedimentary formations comprising weak rock such as shale and stronger rock such as sandstone; these formations often contain vertical stress relief joints. The shale weathers faster than the sandstone, forming overhangs in the sandstone that can have widths of several meters.

Figure 1.2(b) shows a slope in columnar basalt where toppling of the vertical columns is occurring. Toppling, which can be identified by cracks that become narrower with depth, occurs where the base of the column or slab is undermined by weathering of weak seams at the base causing the center of gravity to lie outside the base.

Figure 1.2(c) shows persistent joints dipping out of the face at a dip angle greater than the friction angle of these surfaces, resulting in planar- and wedge-shaped rock falls.

Geology will also influence the shapes of rock falls and their moments of inertia. For example, falls in blocky granitic rock containing orthogonal joints will tend to form cuboid or ellipsoid blocks, while bedded sedimentary rock may form disc-shaped slabs and columnar basalt will form cylindrical columns. In most cases, impacts during the fall will cause the irregular portions of the block to be broken off to form a more uniform shape.

In the failure mechanisms illustrated in Figure 1.2, minor cracks may open on the slope giving an indication of deteriorating stability conditions, but failure usually occurs with no warning. Sudden failure occurs when the stress in the rock exceeds the shear or tensile strength with a significant reduction in strength taking place with even small amounts of movement. Events or conditions that may trigger falls include water or ice pressures acting in cracks, growth of tree roots, and seismic events, as discussed in Sections 1.3 to 1.5.

1.3 WEATHER EFFECTS ON ROCK FALLS

The relationship between rock fall frequency and rainfall and freeze-thaw conditions has been clearly established in previous studies (Peckover, 1975). In another example, Figure 1.3 shows a tabulation of rock fall frequency against monthly rainfall levels and average daily temperature over a period of 20 years for an area on the Pacific Coast of western Canada where the winters are very wet and freeze-thaw cycles occur frequently. Both water and ice

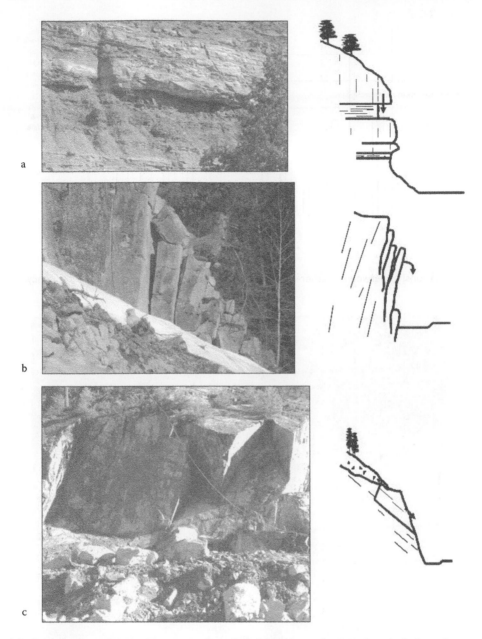

Figure 1.2 Geological conditions that can result in rock falls: (a) weathering of weak shale undercuts beds of strong sandstone forming unstable overhangs (Ohio); (b) toppling columns in columnar basalt (near Whistler, British Columbia); (c) persistent discontinuities dipping out of face allow blocks to slide (near Squamish, British Columbia).

develop pressures in cracks in the rock that can be sufficient to displace and dislodge blocks of rock on the slope face. This is usually a surficial phenomenon because the depth of freezing is limited to 1 to 2 m (3 to 6 ft) behind the face. The most hazardous time is usually when ice melts and releases blocks of rock that have been displaced by the expansion of water when it froze. Figure 1.4 shows a rock slope with heavy water seepage along persistent joints dipping to the right, which has frozen to cover the slope with ice.

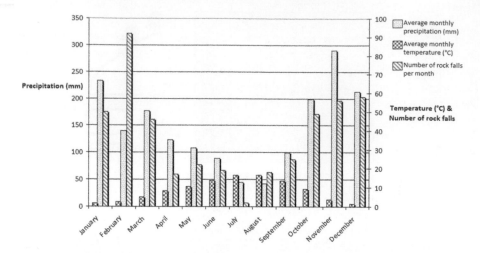

Figure 1.3 Relationship between weather–precipitation and temperature and rock fall frequency.

Figure 1.4 Ice formed on face and in cracks causes loosening of blocks of rock (near Hope, British Columbia). Image by A. J. Morris.

Another study of the causes of rock falls on highways in California showed that nearly two-thirds of rock falls are either directly or indirectly the result of water in the slope (McCauley et al., 1985). Indirect causes of falls related to water include growth of tree roots and weathering of rock.

1.4 VEGETATION EFFECTS ON ROCK FALLS

In wet climates, vegetation and particularly tree growth are usually prolific, and roots can penetrate considerable distances into cracks. The roots are often strong enough to open and extend cracks, and even fracture intact rock. Figure 1.5 shows a root, with a flattened shape, exposed on the face after a rock fall. A further detrimental effect of root growth in cracks is to allow greater penetration of water and ice into the slope that increases rock falls.

1.5 SEISMIC EFFECTS ON ROCK FALLS

Rock falls as the result of ground motions are a common occurrence during earthquakes in mountainous terrain. The events can range from the fall of single rocks, to multiple falls and landslides over wide areas. Kobayashi et al. (1990) describes a block with a mass of 19,000 kg (42,000 lb) that traveled a horizontal distance of 421 m (1,380 ft) from the source on a steep rock face as the result of the magnitude 6.0 earthquake at Mammoth Lakes, California, in 1980. Figure 1.6 shows some of the rock falls resulting from the magnitude 7.9 earthquake in the Denali Mountains in Alaska in 2002 (Harp et al., 2003). This event caused a number of landslides with volumes of tens of thousands of cubic meters on the steep mountain slopes adjacent to the Black Rapids Glacier. Because of the remote location of the earthquake, no damage or injuries resulted from these significant landslides. The

Figure 1.5 Tree root growing in crack in rock has caused rock fall to expose flattened root (near Agassiz, British Columbia).

Figure 1.6 Typical blocks of rock in landslide caused by magnitude 7.9 Denali earthquake in Alaska. Image by Dr. R. Jibson, US Geological Survey.

2011 magnitude 6.2 Christchurch earthquake also caused many rock falls, some of which damaged houses (see Figure 3.1; Dellow et al., 2011).

Harp and Jibson (1995) describe research conducted by the USGS in the Santa Susana Mountains to the east of Los Angeles after the magnitude 6.7 Northridge earthquake in 1994. This event caused approximately 11,000 rock falls and landslides with an average volume of nearly 1,000 cu. m (13,000 cu. yd). This research on the Northridge earthquake and similar events has provided useful guidelines on the relationship between topography, rock strength, ground acceleration, and the risk of seismic slope instability. For example, Keefer (1992) found that the five slope parameters that have the greatest influence on stability during earthquakes are slope angle, weathering, induration, discontinuity characteristics, and the presence of water. The relationship between these parameters and stability is shown on the decision tree in Figure 1.7 that helps to identify site conditions that are susceptible to instability in the event of an earthquake.

The theoretical basis for the occurrence of rock falls and landslides due to ground shaking is Newmark's work on sliding blocks (Newmark, 1965). The Newmark method assumes that the slope comprises a rigid block on a yielding base. When the base moves during an earthquake, displacement of the block will occur if the ground acceleration exceeds the yield acceleration of the block. During the time for each cycle of motion in which the yield acceleration is exceeded, movement of the block will occur, with a cumulative movement occurring over the full duration of the ground motions. Depending on the relative magnitudes of the ground acceleration and the yield acceleration, the block may be displaced, as shown by open tension cracks, but remain on the slope after the ground motions stop, or the block may fall from the slope.

The yield acceleration will be low if the sliding plane is steep and the friction angle is low, compared to a flatter plane and a high friction angle. That is, the Newmark theory demonstrates that rock falls are more likely during an earthquake for blocks on steep, low friction surfaces.

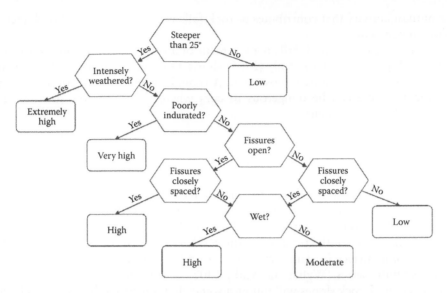

Figure 1.7 Decision tree for susceptibility of rock slopes to earthquake-induced failure. (From Keefer, D. L. 1992. The susceptibility of rock slopes to earthquake-induced failure. *Proc. 35th Annual Meeting of the Assoc. of Eng. Geologists* [ed. Martin L. Stout], Long Beach, CA, pp. 529–538.)

1.6 HUMAN AND ANIMAL INFLUENCES ON ROCK FALLS

Sections 1.2 to 1.5 of this chapter discuss natural factors that can contribute to rock falls. In addition to these factors, human and animal activities can exacerbate rock fall hazards.

The most common cause of rock falls for excavated slopes is blast damage in which excessive blast forces fracture intact rock, displace blocks, and open cracks in the face. Figure 1.8 shows a rock face damaged by blasting—the curved and irregular blast-induced fractures can be distinguished from the natural joints that are planar, which in this case occur in a uniform orientation dipping steeply from right to left on the image.

Opening of cracks in the rock allows water to enter the rock mass, developing pressure in the cracks and possibly ice expansion forces if freezing temperatures occur. In addition, open cracks facilitate the growth of tree roots and further opening of cracks (see Figure 1.5).

Figure 1.8 Blast damage creates open fractures and loose blocks on face.

Another human activity that contributes to rock falls is oversteepening of rock faces resulting in slope movement.

A less common cause of rock falls is movement of animals, such as mountain sheep and goats, on the slope; McCauley's study (1985) showed that 0.3% of rock falls on California highways was the result of animal movement. Although animals usually only dislodge small rock fragments, these can be dangerous to cars where they fall from great heights and impact the road at high velocity.

1.7 CONSEQUENCES OF ROCK FALLS

Rock fall hazards are related to their unpredictability with respect to the source area and the triggering event. That is, blocks of rock that have been observed to be stable for many years may suddenly fall. This is in contrast to soil slides that will usually creep over an extended time before failing. One of the reasons for the sudden failure of rock slopes is that the shear strength of a joint surface in strong rock may drop from peak to residual with a displacement of a few millimetres (Wyllie and Mah, 2002). A further aspect of the hazard is that falls from steep, high rock slopes will follow a somewhat unpredictable path and have a high velocity. A consequence of this combination of hazards is that potentially dangerous falls will occur with no warning. For example, Figure 1.9 shows a car struck by a falling rock that passed through the windshield.

The consequences of rock falls may include both damage and injury such as the event shown in Figure 1.9, as well as delays to traffic, the cost of slope stabilization under costly emergency conditions, and possible legal action.

The design of protection measures against rock falls can take two approaches, depending on the possible consequences of the falls. First, for high-consequence conditions such as high-speed trains and densely populated urban areas, the level of protection provided will ensure that the risk of rock fall accidents is essentially zero. These measures may include the construction of high-energy-capacity rock fall sheds, provision of wide catchment areas, or in extreme cases relocation of transportation systems into fully concrete-lined

Figure 1.9 Windshield of moving vehicle struck by rock fall. Image by N. Boultbee.

tunnels rather than along the base of steep mountain slopes. For example, in the Devils Slide area of the coastal highway south of Pacifica in California, the highway was abandoned after many decades of closures due to rock falls and slides, and relocated into a 2,000 m (6,562 ft) long tunnel.

The second protection approach is to manage the risk where the consequences of falls are less severe such as freight railways and low-traffic-volume roads (Wyllie, 2006). Risk management involves comparing the expected cost of a fall with the cost of providing protection, where the expected cost is the product of the consequence of the fall and the probability of its occurrence. For example, where the most frequent falls are small and large falls occur rarely, adequate protection may be provided by excavating ditches to contain the most frequent falls, or installing catch fences at the base of gullies that are the most common rock fall paths. These protection measures may be designed to contain 85%–95% of the falls, with the understanding that the few falls that exceed the design capacity of the protection have a low probability of causing an accident.

Chapter 8 discusses risk management in more detail, and Chapters 10 and 11 describe a variety of protection measures that range from simple ditches to reinforced concrete rock fall sheds.

Chapter 2

Documentation of Rock Fall Events

This chapter documents five rock fall events that encompass many commonly occurring rock fall conditions. These data are from both natural events where it has been possible to precisely map impact points and trajectories, and from carefully documented, full-scale rock fall tests. The documented events are for a variety of slope geometries and fall heights, and for slope materials comprising rock, colluvium, talus, and asphalt. For these sites, the velocity components in directions normal and parallel to the slope have been calculated from the impact coordinates, and the results have been used to calculate normal and tangential coefficients of restitution, and the energy losses.

The documented events provide reliable data that can be used to calibrate impact and trajectory models. Each of the case studies has been modeled using the program *RocFall 4.0* (*RocScience*, 2012) as described in Chapter 7, where values for the input parameters that are required to fit the calculated trajectories to the field conditions are listed.

Rock falls comprise a series of impacts, each followed by a trajectory, and methods of modeling both impacts and trajectories are required to simulate these events. The basic attributes of trajectories and impacts are as follows:

Trajectory—Rock fall trajectories follow well-defined parabolic paths according to Newtonian mechanics, where three points on the parabola completely define the fall path (Chapter 3). In calculating trajectories at sites where information on precise impact points and trajectory paths is not available, it is necessary to select the two end points for each trajectory and to make an assumption for the angle at which the rock leaves the slope surface. These data have been obtained from measurements at fully documented rock fall sites, and from only using trajectories that are both realistic and mathematically feasible.

Impact—The theory of impact mechanics (Chapter 4) can model rock falls, but has to make simplifying assumptions compared to the actual conditions that occur. Natural conditions include irregularly shaped, translating, and rotating blocks of rock impacting a slope that may be comprised of a different material and also be rough and irregular.

In examining velocity changes during impact, it is useful to calculate the changes in normal and tangential velocity components that occur as the result of deformation and friction at the contact surface. The changes in the velocity components can be quantified in terms of the normal (e_N) and tangential (e_T) coefficients of restitution as defined in the following two equations:

$$\text{Normal coefficient of restitution, } e_N = -\frac{\text{final normal velocity, } v_{fN}}{\text{impact normal velocity, } v_{iN}} \tag{2.1}$$

$$\text{Tangential coefficient of restitution, } e_T = \frac{\text{final tangential velocity, } v_{fT}}{\text{impact tangential velocity, } v_{iT}} \tag{2.2}$$

For each documented rock fall site described in this chapter, insets on the impact drawings show arrows, the lengths and orientations of which are proportional to the velocity vectors. The notation on the vectors include the subscript "i" referring to values at the moment of impact (time, $t = i$), and the subscript "f" refers to values at the end of the impact (time, $t = f$); the final velocity is also referred to as the "restitution" velocity. Also, the subscript "N" refers to the component of the vector normal to the slope, and the subscript "T" refers to the component of the vector tangential to the slope at each impact point. The included angle between the vector and the slope is shown by the symbol θ, with the same subscript designations for impact and final angles.

It is also noted that normal impact velocities ($-v_{iN}$) are negative because the positive normal axis is in the direction out of the slope, and consequentially normal restitution velocities (v_{fN}) are positive. The positive tangential axis is downslope, so all tangential velocities are positive.

This chapter documents actual final velocities and angles measured in the field, while Chapter 3 derives the trajectory equations, and Chapter 4 shows the derivation, based on impact mechanics theory, of equations defining the final velocities and angles. Section 4.7 compares the actual and calculated sets of data for the five documented case studies. Each case study gives the shape, dimensions, mass, and radius of gyration of typical blocks of rock. It has been assumed that the rock fall shapes are either cuboid for falls from low heights, or ellipsoidal where cubic blocks have had the sharp edges and corners broken off by successive impacts on the slope.

2.1 IMPACTS ON ROCK SLOPES

Data have been analyzed for falls at locations in Canada, the United States, and Japan, for slopes ranging in height from 2,000 m to 15 m (6,550 to 50 ft). The following is a discussion on falls at three locations where the falls impacted rock slopes.

2.1.1 Mt. Stephen, Canada—2,000 m High Rock Slope

Mt. Stephen in the Canadian Rocky Mountains near the village of Field is a source of both rock falls and snow avalanches that originate on a rock face with a height of nearly 2,000 m (6,550 ft) at an overall slope angle of about 50 degrees. As shown in Figure 2.1, it has been necessary to construct a barrier to protect a railway operating at the base of the slope. The geology is a strong, tabular, horizontally bedded limestone, containing thin but widely spaced shale beds; the shale weathers more rapidly than the limestone, resulting in the formation of overhangs and falls of the stronger rock.

The barrier comprises a mechanically stabilized earth (MSE) wall built with precast concrete blocks (dimensions 1.5 m long, 0.75 m in section; 5 by 2.5 ft) forming each face, with Geogrid reinforcement and compacted gravel fill between the walls, and a steel cable fence along the top of the wall. The total height of the structure is 11.6 m (38 ft). Figure 2.2 shows a typical section of the lower 120 m (400 ft) of the slope that was generated from an aerial Lidar survey of the site. Figure 2.2 also shows a range of feasible trajectories of rock falls that impacted the lower part of the rock slope and were then contained by the barrier.

It was possible to identify rock fall impact points on both the steel-mesh fence and the concrete blocks and to define the coordinates of each point relative to one end of the wall. In total, 466 impacts were documented. Analyses of typical trajectories that were mathmatically

Figure 2.1 Mt. Stephen rock fall site. (a) View of approximately lower third of rock face with concrete block barrier at base of slope; (b) MSE barrier constructed with concrete blocks, compacted rock fill, and Geogrid reinforcing strips, with steel-mesh fence along top, to contain rock falls and snow avalanches. (Courtesy of Canadian Pacific Railway.)

Figure 2.3 Image of rock fall test carried out in Oregon. (From Pierson, L. A., et al., 2001. Rock Fall Catchment Area Design Guide. Research Report SPR-3(032): Oregon Department of Transportation-Research Group, Federal Highway Administration.)

Figure 2.5 Ehime test site in Japan—rock slope with talus deposit at base; concrete cube test block.

Figure 2.7 Images of Tornado Mountain rock fall. (a) Tree with diameter of about 300 mm (12 in.) sheared by a falling rock at a height of about 1.6 m (5.2 ft); fragment of rock broken off main rockfall visible in lower-left corner; (b) boulder A, with volume of about 1.4 cu. m (1.8 cu. yd), at slope distance of about 740 m (2,450 ft) from source.

Figure 2.9 View of 138 m (450 ft) high slope comprising 58 m (190 ft) high rock slope where rock fall originated, colluvium slope at 42 degrees, and 10 m (33 ft) high-cut face above the road.

Figure 3.1 Examples of impact points visible in the field. (a) Distance successive impact points on slope surface (Christchurch, New Zealand, 2011 earthquake); (b) impact point on tree showing trajectory
height (Tornado Mountain, Canada).

Figure 3.12 Mountain slope with three sinuous gullies in which all rock falls are concentrated.

Figure 6.1 Rock fall that stopped, just before causing serious damage to a building, when it impacted a horizontal surface that absorbed most of the fall energy.

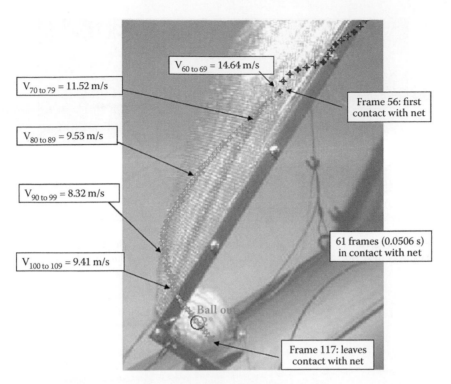

Figure 9.11 Path of deflected projectile after impact with net oriented upslope (β = +60 degrees). Approximate velocities at 10 frame intervals (0.0083 s) during impact shown.

Figure 10.6 Barrier constructed with individual concrete blocks, with five blocks displaced and one fractured by the impact of the rolling boulder shown in the lower image.

Figure 10.7 Rock fall barrier. (a) 8 m (26 ft) high MSE wall with gabions forming facing surfaces, Trans-Canada Highway near Boston Bar, British Columbia. (From British Columbia, MoTI; Maccaferri, 2012.)

Figure 10.12 Slide detector fence for a railway comprising timber posts supporting wires that, if broken by a rock fall, activate signals to stop trains.

Figure 10.20 Typical attenuator hanging net that redirects rock falls into the ditch beside the railway track.

Figure 11.2 Typical precast concrete rock fall shed in Japan. (Courtesy of Protec Engineering, Niigata, Japan.)

a)

b)

Figure 11.7 Examples of cantilevered rock sheds. (a) Cantilevered concrete shed used to deflect rock falls and snow avalanches over railway track. (Courtesy of Canadian Pacific Railway.); (b) "Rock keeper" structure. (Courtesy of Protec Engineering, Japan.)

Figure 11.9 Wire mesh and Ringnet canopy. (Courtesy of Geobrugg, Switzerland.)

Figure 2.1 Mt. Stephen rock fall site. (a) View of lower third, approximately, of rock face with concrete block barrier at base of slope; (b) MSE barrier constructed with concrete blocks, compacted rock fill and Geogrid reinforcing strips, with steel-mesh fence along top, to contain rock falls and snow avalanches. Courtesy Canadian Pacific Railway.

and physically feasible allowed the impact velocity (v_i) and restitution velocity (v_f) to be calculated at each impact point from which the velocity components and tangential (e_T) and normal (e_N) coefficients of restitution were determined. The inset on Figure 2.2 shows the velocity components at impact point A for trajectory $S - A - B$.

The inset shows that velocities at the point of impact for this height of fall can be as great as $30 \text{ m} \cdot \text{s}^{-1}$ (100 ft $\cdot$ s^{-1}). Furthermore, calculation of velocities at the point of impact with the barrier after trajectories that originate at heights of 70 to 100 m (230 to 330 ft) above the barrier can be as high as $48 \text{ m} \cdot \text{s}^{-1}$ (160 ft $\cdot$ s^{-1}). Velocities of this magnitude are consistent with the height of the fall and the steepness of the slope.

The impact energies can be calculated from the mass and velocities of the falls. The rocks tended to break up on impact with the rock slope, and the maximum block dimensions of ellipsoid shaped blocks at the barrier location are about 300 to 500 mm (12 to 20 in), with masses in the range of 50 to 150 kg (110 to 330 lb). Based on a typical velocity at the point of impact with the barrier of about $45 \text{ m} \cdot \text{s}^{-1}$ (150 ft $\cdot$ s^{-1}), the impact energies are approximately 60 to 180 kJ (22 to 66 ft tonf). It was found that the unreinforced concrete blocks

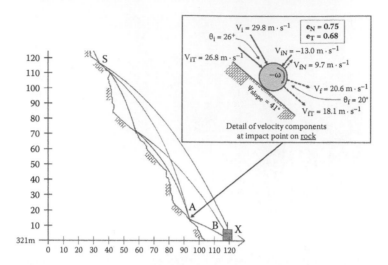

Figure 2.2 Mt. Stephen—cross section of lower part of slope showing ditch and typical trajectories for falls that impact the barrier.

forming the face of the MSE wall were readily able to withstand these impacts, with damage being limited to chips a few millimeters deep.

Analyses of these rock falls using the program *RocFall 4.0* are given in Section 7.5.1.

Typical rock fall properties: ellipsoidal block with axes lengths 0.4 m (1.3 ft), 0.4 m (1.3 ft), and 0.2 m (0.7 ft) mass of 44 kg (97 lb) (unit weight of 26 kN · m⁻³ (165 lbf · ft⁻³)) and radius of gyration of 0.13 m (0.43 ft); (see Table 4.1 for ellipsoid properties).

2.1.2 Kreuger Quarry, Oregon—Rock Fall Test Site

An extensive rock fall test program was carried out at a quarry in Oregon to determine the required ditch configurations to contain rock falls on highways (Pierson et al., 2001). The geometries of the excavated rock cuts included cut heights ranging between 8 m (25 ft) and 24 m (80 ft), face angles ranging between vertical and 45 degrees and ditches inclined at 4H:1V and 6H:1V (toward the slope), and horizontal; in total 11,250 separate rock fall tests were conducted (see Figure 2.3). For each rock fall, the data collected included the first impact position in the ditch, and the roll-out distance. The rock at the test site was a strong, blocky basalt that was excavated with controlled blasting on the design final line to produce a face with few irregularities.

Figure 2.4 shows that test results for a 15 m (50 ft) high cut at a face angle of 76 degrees (¼H:1V) with a horizontal ditch. The diagram shows the measured location of the first impact point in the ditch for the 95th percentile of the test rocks, and the assumed trajectories for a rock fall from the crest of the cut initially impacting the face and then the ditch. The second impact point in the ditch is an estimated location based on common rock fall behavior. The inset on Figure 2.4 shows the calculated velocity components at the impact point on the cut face, and the values for e_T and e_N for the first two impact points. These calculated values show the difference in behavior of rock falls that impact at a shallow angle (on the rock face) and at a steep angle (in the ditch). That is, for shallow angle impacts, the normal coefficient is greater than 1, while for steep angle impacts the normal coefficient is less than 1. These field tests and impact mechanics theory show that e_N is related to the impact angle θ_i, and that e_N is generally greater than 1 when θ_i is less than about 20 degrees (see Figures 5.5 and 5.7). In contrast, the

Figure 2.3 Image of rock fall test carried out in Oregon. (From Pierson, L. A. et al., 2001. Rock Fall Catchment Area Design Guide. Research Report SPR-3(032): Oregon Department of Transportation-Research Group, Federal Highway Administration.)

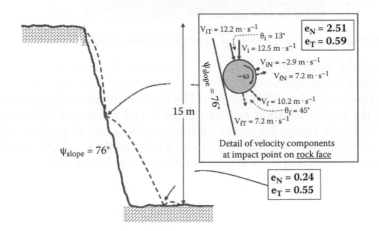

Figure 2.4 Kreuger Quarry, Oregon, test site—typical rock fall trajectory and impact points for 15 m high, 76° rock face with horizontal ditch.

tangential coefficient is less than 1 for both impacts, and the similar values of e_T for the two impacts shown in Figure 2.3 indicate that e_T is a function of the slope material rather than the impact conditions. Coefficients of restitution are discussed in Chapter 5.

Analyses of these rock falls using the program *RocFall 4.0* are given in Section 7.5.2.

Typical rock fall properties: cubic block with side lengths 0.6 m (2 ft), mass of 580 kg (1280 lb) (unit weight of 26 kN · m^{-3} (165 lbf · ft^{-3})), and radius of gyration of 0.245 m (0.80 ft).

2.1.3 Ehime, Japan—Rock Fall Test Site

In 2003, a rock fall study was carried out on a 42 m (140 ft) high rock and talus slope at the Uma-gun Doi-cho test site in Ehime Prefecture on Shikoku Island (Ushiro et al., 2006). The slope comprised a 26 m (85 ft) high rock slope in horizontally bedded sandstone and mudstone with a face angle of 44 degrees, above a 16 m (50 ft) high talus slope at angle of 35 degrees (Figure 2.5).

Figure 2.5 Ehime test site in Japan—rock slope with talus deposit at base; concrete cube test block.

The test involved both natural boulders, and spherical and cubic concrete blocks containing embedded three-dimensional accelerometers and a data acquisition system recording data at a frequency of 1/2000 seconds. The instrumentation together with high-speed cameras gave the precise location and translational and rotational velocities over the full extent of the fall path. Figure 2.6 shows the impact points and trajectories of a typical test of a concrete cube, together with the impact and final velocities, v_i and v_f and the maximum trajectory height h' measured normal to the slope (see also Figure 3.5).

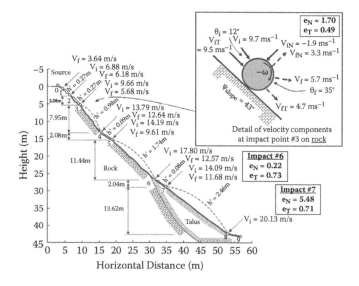

Figure 2.6 Ehime test site, Japan—results of rock fall test showing trajectories, and impact and restitution velocities for concrete cube test block; h' is maximum trajectory height normal to slope, v_i v_f are impact and final velocities. (From Ushiro, T. et al. 2006. An experimental study related to rock fall movement mechanism. *J. Japan Soc. Civil Engineers*, Series F, 62(2), 377–386, in Symp. on Geotechnical and Environmental Challenges in Mountainous Terrain, Kathmandu, Nepal, 366–375.)

Also shown on Figure 2.6 are the calculated velocity components and the values for e_N and e_T at impact point #3 on rock, and the e_N and e_T values for impact points #6 and #7 on talus. For the rock impact, the value for e_N is greater than 1, and has a similar value to that for the Oregon test for the shallow impact angle point on the steep rock face. At both the Oregon and Ehime test sites, e_N values greater than 1 occurred for shallow angle impacts where the impact angles (θ_i) were small (13 and 12 degrees, respectively).

Analyses of these rock falls using the program *RocFall 4.0* are given in Section 7.5.3.

Typical rock fall properties: cubic concrete block with side lengths of 0.6 m (2 ft), mass of 520 kg (1,150 lb) (unit weight of 24 kN · m^{-3} (150 lbf · ft^{-3})), and radius of gyration of 0.245 m (0.80 ft).

2.2 IMPACTS ON TALUS AND COLLUVIUM SLOPES

Information of impacts on talus and colluvium have been obtained from the Ehime test site in Japan (Figure 2.6), and from two rock falls on Tornado Mountain in southeast British Columbia, Canada (Figure 2.8).

2.2.1 Ehime, Japan—Rock Fall Tests on Talus

The calculated e_N and e_T values for impact #6 at Ehime on talus are shown on Figure 2.6. It is of interest that the e_T values of #6 and #7 impacts are nearly identical, while the e_N values are very different; with the trajectory after impact #6 barely leaves the slope surface, while the trajectory after impact #7 is the longest and highest of the rock fall. The difference in the trajectories is probably due to a combination of slope roughness and the attitude of the block as it impacted the surface. The e_N value of 5.48 at impact point #7 is a reliable, measured value that is a significantly higher than other calculated values at this site.

2.2.2 Tornado Mountain—Rock Falls on Colluvium

The Tornado Mountain site comprises a 50 m (165 ft) high rock face in very strong, blocky limestone, above a colluvium slope at an angle varying from 35 degrees on the upper slope in talus to 22 degrees on the lower slope (Figures 2.7 and 2.8). The colluvium is a mixture of gravel and soil forming a uniform slope with no significant irregularities, and no previous rock falls. The slope is sparsely vegetated with pine trees having diameters ranging from about 300 to 500 mm (12 to 20 in.).

In 2004, two separate rock falls originating on the rock face traveled a total distance of 740 m (2,450 ft) down the slope—distances of 340 m (1,115 ft) vertically and 610 m (2,000 ft) horizontally. Because no similar rock falls had occurred in the past and each rock followed a separate path, it was possible to locate each impact point on the slope and define its coordinates with a GPS (global positioning system) unit and a laser rangefinder. In total, 45 impact points were identified for Boulder A and 69 impact points for Boulder B. The final masses for the boulders were about 3,750 kg (8,300 lb; maximum dimension 1.6 m or 5 ft) for Boulder A, and 5,600 kg (12,400 lb; maximum dimension 2.5 m or 8 ft) for Boulder B. Both rocks impacted a horizontal bench in the lower part of the slope that had been excavated in the colluvium beside a railway and the loss of energy on this bench was sufficient to stop the rocks within 30 m (100 ft).

In addition to the impact points on the colluvium, it was also possible to locate a total of 21 trees that had been impacted and broken off by the boulders and measure the height of the impact and the distance of the tree from the two adjacent impact points (see Figure 2.7[a]). It is considered that the trees did not impede the trajectories because of their small diameter

Figure 2.7 Images of Tornado Mountain rock fall. (a) Tree with diameter of about 300 mm (12 in) sheared by
falling rock at a height of about 1.6 m (5.2 ft); fragment of rock broken off main rock fall visible in
lower-left corner; (b) Boulder A, with volume of about 1.4 cu. m (1.8 cu. yd), at slope distance of
about 740 m (2,450 ft) from source.

and low strength. Using information on the coordinates of successive impact points on the
slope and the impact with the trees it was possible for these 21 cases to calculate precise trajectories and velocities, including angles θ_f at the completion of impact and start of the new
trajectory. The average value of θ_f for the tree impacts was 33 degrees, with a range of 6 to
63 degrees (see Figure 3.9[a]); this range of θ_f was entirely due to the variation in the orientation of the blocks of rock at the impact points because the slope surface was uniform. The
average value of θ_f was used to calculate likely velocities components for all other impacts.

Details of the distributions of measured θ_f values at the Tornado Mountain and Ehime
sites are shown in Figure 3.9.

For Boulder A, at impact #26 where the precise trajectory could be determined from a
broken tree and the impact angle θ_i was 22 degrees, the calculated value of e_N is 1.29, a value
that is consistent with other sites where shallow angle impacts occurred. For all 114 impacts
on both fall paths where the paths of the rock falls comprised shallow, "skipping" trajectories, the average value of e_N was 1.02. The calculated value for e_T at impact #A26 was 0.27
as shown on Figure 2.8, with the average value of e_T for Boulders A and B being 0.65.

Figure 2.7(a) shows images of a 250 mm (9.8 in.) diameter tree that was sheared by the rock
fall at a height of about 1.6 m (5 ft) above the ground. Also shown in the image is a block that
broke away from the main block at this impact point. This rock fragment is one of about 20
similar blocks that were observed on the slope over the lower half of the rock fall path; relationships between loss of mass and the run out distance are discussed in Section 6.5.

The trajectories were also analyzed to determine the maximum height of the fall path,
measured normal to the ground surface. It was found that the average height was 1.3 m (4 ft).

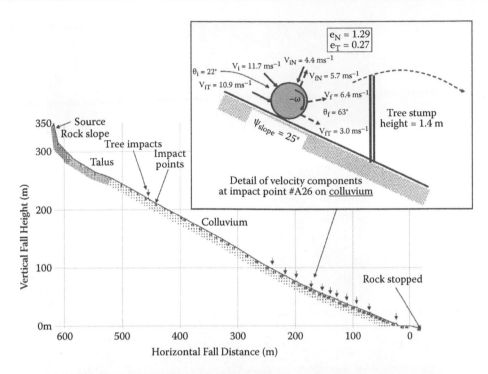

Figure 2.8 Tornado Mountain, Boulder A—mapped impact points (total 46) and broken trees (indicated by arrows ↓), with detail of velocity components at impact #A26.

Analyses of these rock falls using the program *RocFall 4.0* are given in Section 7.5.4.

Typical rock fall properties: ellipsoidal block (Boulder A) with axes lengths 1.6 m (5.2 ft), 1.3 m (4.3 ft) and 1.3 m (4.3 ft), mass of 3750 kg (8300 lb) (unit weight of 26 kN · m⁻³ (165 lbf · ft⁻³)), and radius of gyration of 0.46 m (1.5 ft).

2.3 IMPACT ON ASPHALT

Figure 2.9 shows a 138 m (450 ft) high slope where a single rock fall occurred, that originated at the crest and finally impacted the asphalt road surface. The 138 m (450 ft) high slope comprises a 58 m (190 ft) high rock slope in very strong, volcanic rock at an angle of 60 degrees, a 70 m (230 ft) high colluvium slope at an angle of 42 degrees, and a 10 m (33 ft) high rock cut above the road.

Figure 2.10 shows the final trajectories of the fall from the top of the 10 m (33 ft) high rock cut to just after the impact on the road. These trajectories were precisely defined by a survey of the site, and the inset on Figure 2.10 shows the calculated velocity components at the asphalt impact point. Although this is a single record of an impact with asphalt, the author has investigated several similar events where comparable trajectories were generated for impacts on asphalt.

The trajectory shown in Figure 2.10 is a relatively steep angle impact (i.e., θ_i = 50 degrees) compared to the shallow angle impacts at Tornado Mountain and the Oregon test site, and for this condition angle the value of e_N is 0.38. As discussed in Chapter 5, the value of e_N for steep impacts is low compared to shallow impacts. The value for e_T for this relatively smooth impact surface is 0.24.

Analyses of these rock falls using the program *RocFall 4.0* are given in Section 7.5.5.

Figure 2.9 View of 138 m (450 ft) high slope comprising 58 m (190 ft) high rock slope where rock fall originated, colluvium slope at 42 degrees, and 10 m (33 ft) high cut face above the road.

Typical rock fall properties: cuboid block with axes lengths 0.84 m (2.8 ft), 0.58 m (1.9 ft) and 0.4 m (1.3 ft), mass of 500 kg or 1,100 lb (unit weight of 26 kN · m⁻³ or 165 lbf · ft⁻³), and radius of gyration of 0.295 m (1 ft).

2.4 IMPACT ON CONCRETE

Tests were conducted to find the normal restitution coefficient of concrete. The tests involved dropping a boulder from a known height (h_i) on to a horizontal concrete slab and measuring the rebound height (h_f). It was found that the normal coefficient of restitution e_N for the concrete under these impact conditions was 0.18 $(e_N = (v_f / v_i) = \sqrt{(h_f / h_i)})$—see also Section 5.2.2, Figure 5.6 (Masuya et al., 2001).

2.5 SUMMARY OF CASE STUDY RESULTS

Table 2.1 summarizes the calculated normal and tangential coefficients of restitution for the five case studies, as well as the results of a rock dropped vertically on to a concrete surface. The tabulated results show that e_N has a range of values that are greater than 1 for shallow

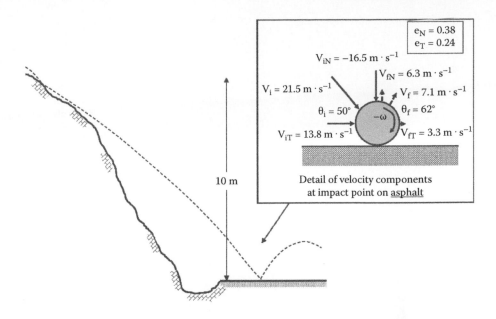

Figure 2.10 Final trajectory of a rock falling from a height of 136 m (445 ft) and impacting a horizontal asphalt surface.

angle impacts and as low as 0.24 for steep angle impacts, and 0.18 for the vertical drop test on concrete; Chapter 5 discusses the relationship between the angle of impact θ_i and e_N.

The results for e_T have a narrower range from 0.25 and 0.27 for relatively smooth asphalt and colluvium surfaces, to 0.49 and 0.73 for rock and talus surfaces. The e_T results show a trend between low values for smooth, soft surfaces to higher values for rough, hard surfaces that is consistent with frictional properties of rock surfaces.

Chapter 4 (impact mechanics) and Chapter 5 (coefficients of restitution) discuss the theoretical relationship between the values of e_N and e_T, and impact conditions. In addition, Section 4.7 compares the final velocities and angles for the five case studies discussed in this chapter, with the calculated final velocities and angles based on impact mechanics theory.

Table 2.1 Summary of coefficients of restitution calculated for rock fall case studies

Site no.	Rock fall site	Slope material	Normal coefficient of restitution, e_N	Tangential coefficient of restitution, e_T
1	Mt. Stephen, Canada	Rock	0.75	0.68
2	Oregon ditch study (rock face impact)	Rock	2.51	0.59
2	Oregon ditch study (ditch impact)	Rock	0.24	0.55
3	Ehime, Japan (rock slope, #3)	Rock	1.70	0.49
3	Ehime, Japan (talus slope, #6)	Talus	0.22	0.73
3	Ehime, Japan (talus slope, #7)	Talus	5.48	0.71
4	Tornado Mountain, Canada	Colluvium	1.29	0.27
5	Highway	Asphalt	0.38	0.24
6	Drop test, Japan (see Figure 5.6)	Concrete	0.18	—

Figure 2.10 Final trajectory of a rock falling from a height of 136 m (445 ft) and impacting a horizontal asphalt surface.

Chapter 3

Rock Fall Velocities and Trajectories

Analysis of rock falls involves study of both trajectories and impacts. This chapter discusses trajectories and how they depend on the translational and angular velocities of the falling bodies. Also discussed are physical characteristics of rock fall sites such as run-out distance, dispersion of falls in the run-out area, and the influence of gullies on rock fall behavior.

Trajectories are defined by the distance between impact points and the height of the rock fall path above the ground surface. In a few instances, it is possible to measure these two parameters in the field, as was the case with the two rock falls at Tornado Mountain discussed in Section 2.2.2, where impacts with trees and the ground could be used to precisely define trajectories. Figure 3.1 shows examples of well-defined impact points.

3.1 TRAJECTORY CALCULATIONS

The trajectories that rock falls follow are exactly defined by Newtonian mechanics, assuming that no air resistance occurs. This section defines the basic equations that govern trajectories and how these are applied to the analysis of rock falls to determine the location of impact points, and the height and length of trajectories.

3.1.1 Trajectory Equation

The trajectory portion of rock falls between impacts, that is, the flight path and changes in the translational velocity, is governed by Newtonian mechanics and gravitational acceleration. The calculations are based on acceleration in the vertical direction being equal to gravity $(a_z = -g)$, with no acceleration in the horizontal direction $(a_x = 0)$. Figure 3.2 shows a body moving with an initial velocity V_0 in direction α_0 relative to a horizontal $[x]$ and vertical $[z]$ coordinate system. Determination of the vertical and horizontal coordinates of the falling rock, and its velocity V_t, at any time t during the trajectory, involves integration between the start of the trajectory at time $t = 0$ and time t, of the following expressions for acceleration and velocity.

Vertical acceleration, $a_z = -g$;   horizontal acceleration, $a_x = 0$

Vertical velocity, $V_{tz} = \int_0^t a_z \, dt$;   horizontal velocity, $V_{tx} = \int_0^t a_x \, dt$

$$= -gt + c \qquad\qquad = c'$$

Figure 3.1 Examples of impact points visible in the field. (a) Distance successive impact points on slope surface (Christchurch, New Zealand, 2011 earthquake); (b) impact point on tree showing trajectory
height (Tornado Mountain, Canada).

At the start of the trajectory when $t = 0$, the velocity components are:

$$V_{tz} = V_{0z} = c \quad \text{and} \quad V_{tx} = V_{0x} = c'$$
$$V_{tz} = -gt + V_{0z} \qquad V_{tx} = V_{0x}$$

Vertical location, $z = \int_0^t (-gt + V_{0z})\,dt$;   horizontal location, $x = \int_0^t (V_{0x})\,dt$ (3.1)

$$z = -\frac{1}{2}gt^2 + V_{0z}t + c'' \qquad\qquad x = V_{0x}t + c'''$$

When $t = 0$, $z = 0$, $x = 0$, and $c'' = c''' = 0$

Therefore, $z = -\dfrac{1}{2}gt^2 + V_{0z}t$; $x = V_{0x}t$ or $t = x/V_{0x}$ (3.2)

where V_{0z} and V_{0x} are, respectively, the velocity components in the vertical and horizontal
directions at the start of the trajectory.

Equation (3.2) defines a parabolic curve that gives the $[x,z]$ coordinates of rock fall trajectories and other projectiles, as a function of time of flight, t.

The angle α_0, measured counterclockwise from the positive x-axis, defines the direction
of the velocity vector V_0 at time $t = 0$ (Figure 3.2) from which the following expressions can
be obtained:

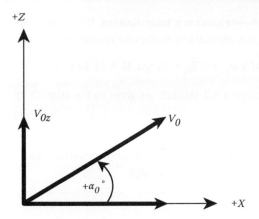

Figure 3.2 Definition of trajectory velocity components and directions.

$$\tan\alpha_0 = \frac{V_{0z}}{V_{0x}} \quad \text{and} \quad V_{0x} = V_0 \cdot \cos\alpha_0 \tag{3.3}$$

Equations (3.2) and (3.3) can be combined to define the location of the body in terms of $[x, z]$ coordinates, the initial velocity V_0, and angle α_0 as shown in Equation (3.4):

$$z = -\frac{g}{2}\left(\frac{x}{V_0\cos\alpha_0}\right)^2 + x\tan\alpha_0 \tag{3.4}$$

Equation (3.4) can be rearranged to show the relationship between the initial velocity V_0, of the body, its angle relative to the x-axis α_0, and the distance traveled by the body from the initial point to the point defined by the coordinates $[x, z]$:*

$$V_0 = \frac{x}{\cos\alpha_0\left[\dfrac{2(x\cdot\tan\alpha_0 - z)}{g}\right]^{0.5}} \tag{3.5}$$

---

* Equation (3.5) can also be used for applications other than rock falls. For example, for water discharging from a pipe, the vertical and horizontal distances of the jet from the discharge point can be measured as well as the angle of the pipe. These values can be entered in Equation (3.5) to find the discharge velocity, from which the flow rate can be calculated knowing the discharge area.

Worked example 3A—trajectory coordinates: If the initial velocity, $V_0 = 20$ m · s^{-1} at an angle $\alpha_0 = 35$ degrees, the vertical and horizontal components of the initial velocity are

$$V_{0z} = 20 \sin 35 = 11.5 \text{ m} \cdot \text{s}^{-1}; V_{0x} = 20 \cos 35 = 16.4 \text{ m} \cdot \text{s}^{-1}$$

and the coordinates at time $t = 2$ seconds are given by Equation (3.2):

$$z = -(0.5) \cdot (9.81) \cdot (2)^2 + (11.5) \cdot (2) = 3.38 \text{ m}$$

If $t = 4$ seconds, $z = -32.5$ m, showing that, at this time when $x = 65.6$ m, the body has descended below the x-axis and now has a negative z value.

3.1.2 Nomenclature—Trajectories and Impacts

The nomenclature used to define the velocities and angles of trajectories and impacts is shown in Figure 3.3. For trajectories, the start of the trajectory is designated by the subscript "0" (time, $t = 0$) and the end of the trajectory where the next impact starts is designated by the subscript "i" (time, $t = i$).

With respect to the impact, the moment of impact at the end of the trajectory is designated by the subscript "i" (e.g., V_i, θ_i), and the end of the impact and the start of the next trajectory by the subscript "f" (e.g., V_f, θ_f). That is, $V_f = V_0$ and $\theta_f = \theta_0$, and the velocity at the end of the impact equals the velocity at the start of the next trajectory. The final velocity and angle are also termed the "restitution" parameters. Chapter 4, Impact Mechanics, discusses methods of calculating velocity components during impact.

3.1.3 Rock Fall Trajectories

The trajectory equations discussed in Section 3.1.1 can be applied directly to the analysis of rock falls to determine parameters such as impact and restitution velocities, and trajectory lengths and heights.

Figure 3.3 shows a typical rock fall configuration at the completion of an impact where a body is just leaving the ground (impact point, (n)) at a final (restitution) velocity V_f, equal to the velocity at the start of the trajectory, V_0. The orientation of this velocity vector can be defined by either of two angles, α_0 or θ_0. The angle α_0 is relative to the positive x-axis, measured either counter-clockwise (positive) or clockwise (negative) and can have values between 0 and 360°. Angle α_0 is used for the application of Equations (3.1) to (3.5); in applying these equations, values of α_0 of +330 degrees or −30 degrees, for example, give identical results. That is, $\alpha_0 = -30$ degrees is equivalent to $\alpha_0 = +330$ degrees because $\sin(-30) = \sin(330) = (-0.5)$.

The angle θ_0, at the start of the trajectory, is used to define the orientation of the velocity vector relative to the ground surface and is a parameter that is more readily used in the analysis of field data. If the slope angle is ψ_s, then the relationship between these three angles is given by:

$$\theta_0 = (\psi_s + \alpha_0) \tag{3.6}$$

Equation (3.6) is applied for positive and negative values of α_0.

An essential premise of Newtonian mechanics applied to rock falls is that while the vertical component of the translational velocity changes during the trajectory as a result of

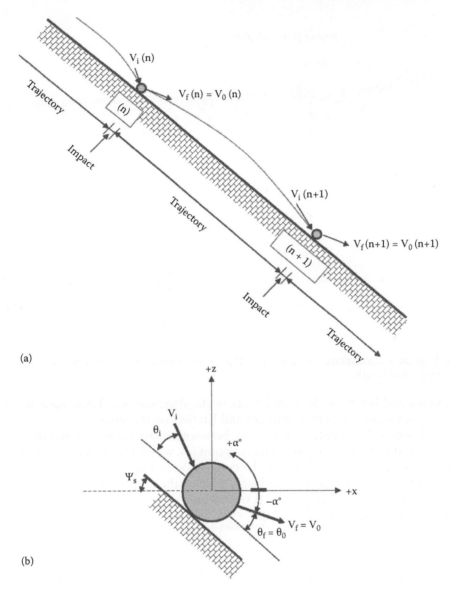

Figure 3.3 Definition of trajectory parameters. (a) Velocity nomenclature for trajectories and impacts; (b) parameters used in Equation (3.4) to calculate rock fall trajectories.

gravitation acceleration, both the horizontal velocity and the angular velocity are constant during the trajectory because no forces act on the body to change these velocity components.

The full trajectory of the rock fall, defined by the $[x, z]$ coordinates, can be obtained from Equation (3.4) for specified values of V_0, θ_0, and ψ_s. Figure 3.4 shows successive locations of a rock fall following a parabolic trajectory between impact points (n) and $(n+1)$.

3.1.4 Trajectory Height and Length

The trajectory equations discussed in Section 3.1.1 above that define the rock fall path through the air, can also be used to find the next impact point, the slope distance between

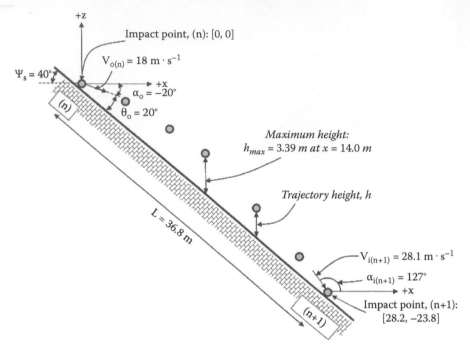

Figure 3.4 Trajectory calculations showing rock fall path, impact points, impact velocities, and trajectory height and length.

impact points, and height of the rock fall above the slope surface. These data are useful in designing the location and height of fences and barriers on the slope.

Figure 3.4 shows the calculated trajectory between impact points (n) and $(n + 1)$ for the velocity vector and slope angle parameters at point (n), where the velocity at the start of the trajectory V_0 is equal to the velocity at the completion of impact at point (n), or $V_{f(n)}$. The coordinates of impact point $(n + 1)$ can be found from the point of intersection between the equations for the trajectory and slope. If the average slope between the impact points has gradient K, then the equation of the slope is

$$z = K x \tag{3.7}$$

and the point of intersection is found by equating Equations (3.4) and (3.7) as follows:

$$K x = -\frac{g}{2}\left(\frac{x}{V_0 \cos\alpha}\right)^2 + x \tan\alpha$$

and

$$x_{impact} = \left[\frac{-2(K - \tan\alpha)}{g\left(\dfrac{1}{V_0 \cos\alpha}\right)^2}\right] \tag{3.8}$$

Once the x coordinate of the impact point is known, it can be substituted in either Equation (3.4) or (3.7) to find the z coordinate.

The trajectory equation can also be used to find the vertical height h of the body above the slope at any point, as well as the maximum height h_{max}, and the x coordinate of this height. The height of the body above the slope is equal to the difference in z coordinates given by equations (3.2) and (3.7).

$$h = (z_{traj} - z_{slope}) \tag{3.9a}$$

$$= (ax^2 + bx) - (Kx)$$

$$= ax^2 + x(b - K)$$

$$\text{where } a = -\frac{g}{2}\left(\frac{1}{V_0 \cdot \cos\alpha}\right)^2 \text{ and } b = \tan\alpha \tag{3.9b}$$

Equations (3.9a) and (3.9b) define the height of the body above the slope for any value of the horizontal coordinate [x].

The maximum height of the body above the slope can be found by differentiation of Equation (3.9a) and equating the result to zero. The differential of Equation (3.9a) is

$$dh/dx = 2ax + (b - K) \tag{3.10}$$

and the value of the x coordinate where the height is a maximum is

$$x = \frac{-(b - K)}{2a} \tag{3.11}$$

This value of x can then be substituted in Equation (3.9a) to calculate the maximum trajectory height, h_{max} (Figure 3.4).

Worked example 3B—trajectory calculations: a rock fall comprises a series of impacts with a slope at a uniform angle ψ_s = 40 degrees (gradient, K = −0.84). At impact point (n) with coordinates [0, 0], the velocity at the start of the trajectory is $V_{0(n)}$ = 18 m · s^{-1} (Figure 3.4), and the angle of the velocity vector at this point is $\theta_{0(n)}$ = 20 degrees. The angle α_0 defining the orientation of the velocity vector can be found from Equation (3.6) to be α_0 = −20 degrees.

Based on these parameters, the coordinates of any point on the parabolic trajectory can be found from Equation (3.4):

$$z = -0.017x^2 - 0.36x$$

The equation of the slope is (z = −0.84x), so the coordinates of the next impact point (n + 1) can be found from equating these two expressions for z. The coordinates of impact point (n + 1) are: [$x_{(n+1)}$ = 28.2 m, $z_{(n+1)}$ = −23.8 m].

The slope length L between these two impact points is obtained from the difference Δ, of their vertical (z) and horizontal (x) coordinates:

$$L = (\Delta x^2 + \Delta z^2)^{0.5}$$

$$= 36.8 \text{m}$$

From Equation (3.11), the maximum vertical height of the trajectory above the slope is 3.39 m, which occurs at x coordinate 14.0 m.

It is also possible to calculate the impact velocity at impact point $(n + 1)$, $V_{i(n+1)}$ as follows. The horizontal and vertical components of the velocity at impact point (n) at the start of the trajectory are, respectively:

$$V_{0x(n)} = V_{0(n)} \cos \alpha = 16.9 \text{ m} \cdot \text{s}^{-1} \text{ and } V_{0z(n)} = V_{0(n)} \sin \alpha = 6.2 \text{ m} \cdot \text{s}^{-1}$$

During the trajectory to impact point $(n + 1)$, the vertical velocity component will increase due to gravitational acceleration and the vertical velocity at point $(n + 1)$ is

$$V_{iz(n+1)} = \sqrt{(V_{0z(n)})^2 + 2g(\Delta z)}$$

$$= 22.4 \text{ m} \cdot \text{s}^{-1}$$

The horizontal velocity component will not change during the trajectory so that the two components of the impact velocity at impact point $(n + 1)$ are $V_{ix(n+1)} = 16.9 \text{ m} \cdot \text{s}^{-1}$ and $V_{iz(n+1)} = 22.4 \text{ m} \cdot \text{s}^{-1}$. The resultant impact vector at point $(n + 1)$ has a velocity of $V_{i(n+1)} = 28.1 \text{ m} \cdot \text{s}^{-1}$. The angle $\theta_{i(n+1)}$ of the vector is given by:

$$\theta_{i(n+1)} = atan\left(\frac{22.4}{16.9}\right) = 53 \text{ degrees}$$

at an angle of $\alpha_{i(n+1)} = 180 - 53 = 127$ degrees.

3.1.5 Field Trajectory Heights

Section 3.1.4 above and Figure 3.4 show the calculated theoretical trajectory height measured vertically from the slope surface for a parabolic rock fall trajectory. Information on actual trajectory heights is available from the Ehime test site in Japan where the test blocks comprised concrete spheres with a diameter of 0.54 m (1.8 ft), concrete cubes with a side length of 0.6 m (2 ft), and blocks of rock with masses ranging from 120 kg to 2,060 kg (260 lb to 4550 lb). The test program consisted of 10 spheres, 10 cubes, and 20 blocks of rock, with half of the tests carried out on a treed slope and the second half after the trees had been removed (see Section 6.6 regarding the energy dissipation of trees growing in the rock fall path). The concrete bodies contained embedded three-dimensional accelerometers that provided detailed information on the positions and velocities of the blocks throughout the fall. The slope at the test site was 42 m (140 ft) high, made up of a 26 m (85 ft) high sandstone, and mudstone rock slope with a face angle of 44 degrees, and a 16 m (50 ft) high talus cone at an angle of 35 degrees (see Section 2.1 and Figure 2.6).

A component of the data collected was the heights of the trajectories above the slope surface, in a direction at right angles to the slope. These data are shown on Figure 3.5 where the height is plotted against the fall height from the source. The results show that the heights vary from zero, i.e., a rolling block, to a maximum height of about 2 m (6.5 ft), with no significant difference in the trajectory heights between the three block shapes. Further analysis of the data shows that, for a total of 235 trajectories, 233 (or 99%) had heights of less than 2 m (6.5 ft), and that 56% had heights less than 0.5 m (1.6 ft). For the 11 trajectories (4.7%)

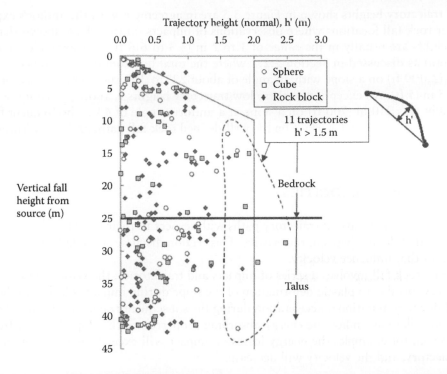

Figure 3.5 Plot of normal trajectory heights from Ehime test site for spherical and cubic concrete blocks, and blocks of rock (see Figure 2.6 for slope section). (From Ushiro, T. et al. 2006. An experimental study related to rock fall movement mechanism. *J. Japan Soc. Civil Engineers*, Series F, 62(2), 377–386 in Symp. on Geotechnical and Environmental Challenges in Mountainous Terrrain, Kathmandu, Nepal, 366–375.)

where the heights exceeded 1.5 m (5 ft), the preceding impact involved a projecting rock or tree that deflected the fall away from the slope.

It is noted that, while Equations (3.9a) and (3.9b) define the vertical height of the trajectory above the slope surface, the data shown in Figure 3.5 is for the height (h') normal to the slope. The normal trajectory height is defined as follows:

$$h' = \frac{(V_0 \cdot \sin\theta_0)^2}{2g\cos\psi_s}$$

(3.12)

For the rock fall parameters used in Worked example 3B, the maximum normal trajectory height is:

$$h' = \frac{(18\sin[20])^2}{2g\cos(40)} = 2.52 \text{ m}$$

This maximum height in the normal direction compares with the maximum vertical height of 3.38 m.

In the design of nets and barriers it is often acceptable to provide protection for about 90 to 95% of the falls. Under these conditions, the height of a net at Ehime, for example, would need to be 1.5 m (5 ft) high to contain 95% of the falls (Figure 3.5).

The trajectory heights shown in Figure 3.5 are in agreement with the author's experience of other rock fall locations where observations of impacts on trees have shown that trajectory heights are usually in the range of 1 to 2 m (3.3 to 6.6 ft). For example, at Tornado Mountain as discussed in Section 2.2.2 where the total horizontal fall distance was about 610 m (2,000 ft) on a slope with an angle of about 22° to 30°, the average trajectory height was 1.5 m (5 ft). An exception to these low trajectory heights is another location where the fall height was 210 m (700 ft) on a slope at a uniform angle of 43°. The frequent falls had produced a polished rock surface on the well-defined rock fall path, and in the lower part of the slope, trajectory heights of up to 4 m (13 ft) were observed.

3.2 ROCK FALL VELOCITIES

Section 3.1 above discussed trajectory paths with respect to their height and length and the equations that define these characteristics. This section discusses rock fall velocities and the conditions that influence velocity.

When a rock fall involves a series of impacts and trajectories, the velocity will increase if the energy lost due to plastic deformation of the slope during impact is less than the energy gained due to gravitational acceleration during the subsequent trajectory. Eventually, as the slope angle flattens and/or the energy absorbing properties of the slope increase from bare rock to soil, for example, the energy losses on impact will exceed the energy gains during the trajectory, and the velocity will decrease.

3.2.1 Field Velocity Measurements

Typical velocities for the Ehime test site in Japan, as a function of the fall height, are shown in Figure 3.6, in which the impact (V_i) and restitution (V_f) velocity components are shown separately (see Figure 2.6 for slope section). The plot shows that impact velocities are greater than restitution velocities, representing the loss of energy at the impacts; the variation in the velocities is the result of the details of differing geometric conditions at the impact points.

Features of the plot in Figure 3.6, which are typical of rock falls, are that the velocities increase in the early part of the fall over a height of about 15 m (50 ft) in this case, and then reach an approximate terminal velocity of about 15 to 18 m · s⁻¹ (50 to 60 ft · s⁻¹). The plot also shows that velocities in the rock and talus portions of the slope, with the velocities on the talus being somewhat less than those on the rock. The velocities decrease at fall heights greater than 40 m (130 ft) where the rocks roll past the base of the talus slope on to flatter ground and stop moving.

3.2.2 Effect of Friction and Slope Angle on Velocity

The main site characteristics that influence rock fall velocities, in addition to the fall height, are the slope angle and the material(s) on the slope surface, with lower velocities for shallow slopes and softer, more energy-absorbing material compared to higher velocities for steep, hard rock slopes. The relationship between velocity and height, incorporating the slope angle and the characteristics of the slope material, is discussed below.

Referring to Figure 3.7(a), a block of rock with mass m, is sliding on a slope dipping at ψ_s degrees with the surficial material having an effective friction coefficient μ'. The motion of this block can be studied using limit equilibrium methods that compare the relative magnitudes of the driving and resisting forces (Wyllie and Mah, 2002). If the driving force is

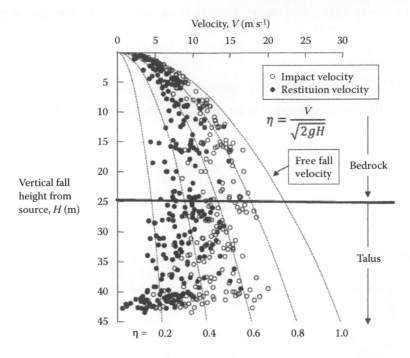

Figure 3.6 Range of velocities for Ehime rock fall test site. (See Figure 2.6 for slope section.)

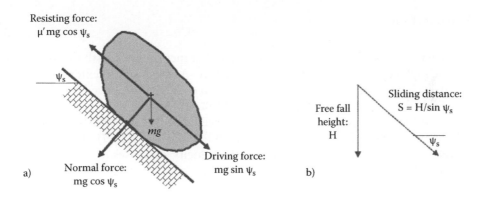

Figure 3.7 Velocity of rock fall on slope dipping at ψ_s: (a) limit equilibrium forces acting on sliding block; (b) relationship between free fall height, H, and sliding distance, S.

greater than the resisting force, then the out-of-balance force $(m \cdot a)$ causes acceleration a of the block according to the following relationship:

$$m \cdot a = (\text{driving force} - \text{resisting force})$$

$$= (m \cdot g \cdot \sin \psi_s - \mu' \cdot m \cdot g \cdot \cos \psi_s)$$

$$a = g \cdot \sin \psi_s \left(1 - \frac{\mu'}{\tan \psi_s} \right)$$

Referring to Figure 3.7(b), the velocity V of the block sliding down the sloping plane over a distance S is:

$$V^2 = 2aS = \frac{2aH}{\sin \psi_s}$$

and the rock fall velocity expressed in terms of the vertical fall height H is:

$$V = \left[2gH \left(1 - \frac{\mu'}{\tan \psi_s} \right) \right]^{0.5} \tag{3.13}$$

Equation (3.13) can also be expressed in the following manner:

$$\eta = \frac{V}{[2gH]^{0.5}} \tag{3.14}$$

where η is a parameter representing the slope characteristics:

$$\eta = \left(1 - \frac{\mu'}{\tan \psi_s} \right)^{0.5} \tag{3.15}$$

or

$$\mu' = \tan \psi_s (1 - \eta^2) \tag{3.16}$$

The parameter μ' is an effective friction coefficient that incorporates both the material forming the slope surface and the roughness of this surface. As the result of extensive field testing of rock falls in Japan, values for the effective friction coefficient μ' have been determined as shown in Table 3.1 (Japan Road Association, 2000).

Referring to Figure 3.6, the series of dashed curves show the relationship between values of η and the measured rock fall velocities. When $\eta = 1$ and $V = \sqrt{(2gH)}$, the curve shows the free-fall velocity. For values of η less than 1, the velocity decreases as shown by the set of curves for η values of 0.8 to 0.2.

Table 3.1 Values of effective friction coefficient μ' for characteristics of slope materials

Slope category	Characteristics of slope surface materials	Design μ' values	Range of μ' from field tests
A	Smooth, strong rock surfaces and uniform slope profile; no tree cover	0.05	0.0 to 0.1
B	Smooth to rough, weak rock surfaces with medium to high roughness slope profile; no tree cover	0.15	0.11 to 0.2
C	Smooth to rough, weak rock, soil, sand, or talus with low to medium roughness slope profile; no tree cover	0.25	0.21 to 0.3
D	Talus with angular boulders exposed at surface, medium to high roughness slope profile; no tree cover or few trees	0.35	~0.31

*These values for μ' tend to give upper bound velocity values.

Worked example 3C—fall velocities: the Ehime test site has two slope components: a 26-m-high rock slope at an average angle of 44 degrees, and a 16-m-high talus slope at an angle of 35 degrees (see Sections 2.1 and 2.2, and Figure 2.6). From Table 3.1, the approximate friction coefficients are 0.15 for the rock and 0.35 for the talus.

Figure 3.6 shows the measured velocities at the site, together with curves representing values for the parameter η. On Figure 3.5, the maximum velocities correspond to values of η of $\eta_{rock} = 0.9$ and $\eta_{talus} = 0.7$, Equation (3.16) gives values for the friction coefficients as follows:

$$\mu'_{rock} = \tan 44 \ (1 - 0.9^2) = 0.18 \text{ and for talus, } \mu'_{talus} = \tan 35 \ (1 - 0.7^2) = 0.36$$

Comparison of these calculated values for μ' with the values shown on Table 3.1, shows that equation (3.13) gives values for the maximum velocities if the friction angles given in Table 3.1 are applied.

In another example, at Tornado Mountain (see Section 2.2.2 and Figure 2.8) where the slope has low roughness comprising gravel and soil, the impact velocities are in the range of 10 to 15 $m \cdot s^{-1}$ on the lower part of the slope where the slope angle is 22 degrees and the fall height H, is about 250 m. The value for η is given by (equation (3.14):

$$\eta = \frac{12.5}{(2 \cdot g \cdot 250)^{0.5}} = 0.39 \quad \text{and} \quad \mu'_{colluvium} = \tan 22 (1 - 0.39^2) = 0.34$$

This value for $\mu'_{colluvium}$ is consistent with values for the friction coefficient given in Table 3.1 for talus.

These back analyses for the relationship between velocity, slope angle, and friction show that the friction coefficients given on Table 3.1 can be used to estimate velocity values, although these velocities may be at the high end of the actual velocity values.

The relationships in Equations (3.12) to (3.14) and the effective friction coefficients shown in Table 3.1 are used in Japan to estimate fall velocities and impact energies in the design of protection structures. The application of these relationships to the Ehime and Tornado rock fall sites illustrates that this is a useful method to estimate rock fall velocities when no field data are available.

3.3 VARIATION OF TRAJECTORIES WITH RESTITUTION ANGLE

This section discusses the variation of the angle at the start of the trajectory θ_0, following an impact, and its influence on the length and height of trajectories. This can be an important factor in the design of protection structures that should be positioned at low-trajectory and low-velocity locations on the slope.

3.3.1 Calculated Trajectories for Varying Restitution Angles (θ_0)

Figure 3.8 shows two possible trajectories between a pair of impact points, (n) and $(n + 1)$ on the slope dipping at angle $\psi_s = 30$ degrees, together with the initial velocities and angles at point n. At the start of the trajectory, the body leaves the slope with a restitution velocity V_0, and at angle θ_0 relative to the slope surface. The $[x, z]$ coordinates of the body at any point during the trajectory can be calculated using Equation (3.4). Figure 3.8 shows the successive positions of the body for two possible trajectories defined by the initial angles $\theta_0 = 15$ degrees

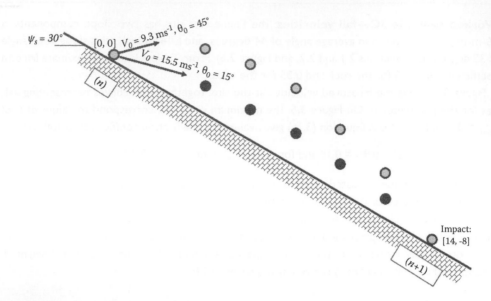

Figure 3.8 Trajectories related to restitution angle, θ_0: $\theta_0 = 15°$ and $\theta_0 = 45°$.

and $\theta_0 = 45$ degrees. In applying Equation (3.4) to calculate the $[x, z]$ coordinates of the trajectories, the α_0 angles for the upper and lower trajectories are, respectively, $\alpha_0 = +15$ degrees and $\alpha_0 = -15$ degrees, where α_0 is defined by Equation (3.6) – ($\alpha_0 = \theta_0 - \psi_s$).

Also, Equation (3.5) can be used to calculate the velocity required for the body to reach the impact point (point $(n + 1)$) with coordinates: $[x_i = 14]$ and $[z_i = -8]$. For the low-angle trajectory ($\theta_0 = 15$ degrees), the required velocity is 15.5 m · s^{-1} (50.9 ft · s^{-1}), while for the higher angle trajectory ($\theta_0 = 45$ degrees), the required velocity is only 9.3 m · s^{-1} (30.5 ft · s^{-1}). That is, as the height of the trajectory increases, the velocity required for the body to reach a defined point decreases.

3.3.2 Field Values of Restitution Angles (θ_0)

Studies of the rock falls at Tornado Mountain and Ehime in Japan discussed in Chapter 2 have been carried out to investigate the variation in restitution angles θ_0 that occurs in the field (Figure 3.9(a) and (b)).

For the two Tornado Mountain rock falls (see Section 2.2.2), the coordinates for a total of 114 impact points were measured in the field from which it was possible to determine the slope length and inclination of each trajectory. However, it was only possible to determine the trajectory path from Equation (3.4) if information was also available on the magnitude of the angle α_0 defining the inclination of the initial velocity relative to the x-axis (Figure 3.2). This information was provided at 21 locations along the trajectories where rock falls impacted trees, and it was possible to measure the height of the impact, and the distance of the tree from the preceding impact point. These measurements provided three points on the parabolic path of the rock fall from which the trajectory parameters, including the value of the angle α_0, could be precisely calculated.

From the 21 calculated values of α_0 angles and the measured slope angle ψ_s between the two impact points on the slope, values of the initial angle θ_0 were calculated using Equation (3.6). It was found that the values of θ_0 ranged from 6° to 63°, a difference of 57°, with an average value of 33° (Figure 3.9a). That is, the impact process caused the restitution angle to

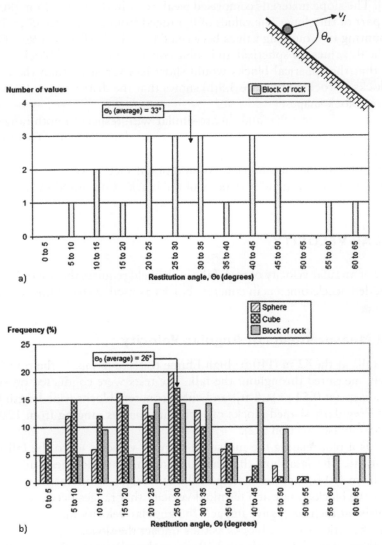

Figure 3.9 Ranges of values for restitution angle, θ_0. (a) Tornado Mountain tree impacts (21 points); (b) Ehime test site trajectory measurements for spherical, cubic concrete blocks and blocks of rock. (From From Ushiro, T. et al. 2006. An experimental study related to rock fall movement mechanism. *J. Japan Soc. Civil Engineers*, Series F, 62(2), 377–386, in Symp. on Geotechnical and Environmental Challenges in Mountainous Terrrain, Kathmandu, Nepal, 366–375.)

vary from a shallow angle in which the block barely left the slope surface, to larger angles where the block follows a relatively high trajectory. Analysis of trajectories shows that the height of the trajectory does not significantly influence the distance between impact points because the restitution velocity decreases as the trajectory height increases, i.e., high angle trajectories have velocities that are less than shallow angle trajectories.

A particular feature of the Tornado Mountain site is the uniform slope, at an angle of between 22 to 30 degrees, composed of gravel and soil with essentially no surface roughness or irregularities. Therefore, variation in the restitution angle would be entirely the result of the attitude of the rotating, irregular block as it impacted the slope.

Values of restitution angles were also measured at the Ehime test site (see Sections 2.1 and 2.2) using data from the embedded accelerometers and the high-speed cameras

(Figure 3.9b). The slope materials comprised weak rock in the upper 25 m (80 ft) and talus in the lower part of the slope. The values of θ_0 ranged from about 5 to 55°, with an average of 26°, discounting two outlying values between 65° and 70° (Figure 3.9b). The histogram also shows the θ_0 values for spherical and cubic concrete blocks and blocks of rock. It may be expected that the spherical blocks would show less variation than the more irregular cubes and blocks of rock, but Figure 3.9(b) shows that the distribution of θ_0 values is similar for all three block shapes.

The two plots in Figure 3.9(a) and (b) are similar with respect to both range and form of the distributions, despite the significantly different site characteristics. This range of θ_0 values accounts for the variation of trajectories that are usually observed in the field, even where conditions are similar for each impact. An additional factor influencing variations in the values of the angle θ_0 is the angular velocity of the block as discussed in Section 3.4 below.

3.4 ANGULAR VELOCITY

Measurement of angular velocity of rock falls in the field requires the use of high-speed cameras or embedded accelerometers in concrete blocks as used at the Ehime test site in Japan.

3.4.1 Field Measurements of Angular Velocity

For the rock falls at the 42 m (140 ft) high Ehime test site in Japan, the rotational velocity was accurately measured throughout the fall. The tests were conducted on spherical concrete blocks (diameter 0.54 m [1.8 ft]) and cubic concrete blocks (side length 0.6 m [2 ft]), as well as 20 irregularly shaped blocks of rock with masses ranging from 120 to 2,060 kg (260 lb to 4,550 lb) (see Section 2.2).

Figure 3.10 is a plot relating the measured angular velocity, ω, to the fall height, H for the Ehime test site. The data show that the range of angular velocities was 6 to 33 rad · s^{-1}. For the first 10 m (30 ft), approximately of the fall height, ω, increases with each impact, and thereafter the blocks rotate at a terminal velocity that ranges between about 12 and 30 rad · s^{-1}. That is, at each successive impact, the angular velocity may increase or decrease, depending on the attitude of the blocks as they impact the slope.

Further analysis of the data on Figure 3.10 shows the relationship between the dimensions and shapes of the blocks and the angular velocity. The range of angular velocities shown in Figure 3.10 is similar for all three block shapes, with the spherical and cubic concrete blocks reaching slightly higher angular velocities of about 32 rad · s^{-1} than that for the irregular blocks of rock at 27 rad · s^{-1}.

The angular velocity measurements have also been analyzed to show the relationships between the translational and angular velocities at the start of the trajectory V_0 and ω_0, respectively, of the blocks and their dimensions r. The theoretical relationship between these three parameters is as follows:

$$\omega_0 = \frac{V_0}{r} \quad \text{or} \quad r = \frac{V_0}{\omega_0} \tag{3.17}$$

For the three types of blocks used at the Ehime test site, the following values were obtained for values the radius r, as defined by Equation (3.17).

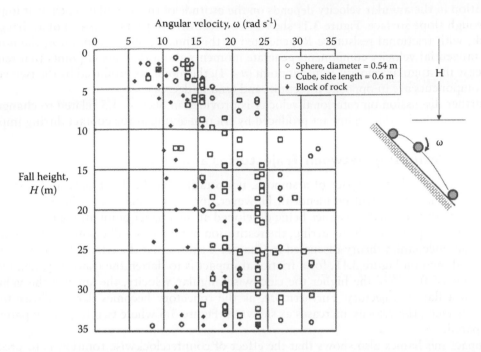

Figure 3.10 Relationship between angular velocity and fall height for rock falls at Ehime test site, Japan for spherical and cubic concrete blocks, and of blocks of rock. (From Ushiro, T. et al. 2006. An experimental study related to rock fall movement mechanism. *J. Japan Soc. Civil Engineers*, Series F, 62(2), 377–386, in Symp. on Geotechnical and Environmental Challenges in Mountainous Terrrain, Kathmandu, Nepal, 366–375.)

- Spheres with radius, $r = 0.27$ m,

$$\frac{V_0}{\omega_0} = 0.2 \text{ to } 0.54 \text{ m}$$

- Cubes with side length 0.6 m and semi-diagonal length, $r = 0.42$ m,

$$\frac{V_0}{\omega_0} = 0.25 \text{ to } 0.7 \text{ m}$$

- Blocks of rock with an average radius, $r = 0.65$ m,

$$\frac{V_0}{\omega_0} = 0.3 \text{ to } 1.0 \text{ m}$$

For all three types of block, the theoretical relationship defined by Equation (3.17) is reasonably consistent with the field values. That is, the actual radius of the rotating body lies within the range of field values for the ratio (V_0/ω_0). Also, the tests showed that the range of field values for the ratio (V_0/ω_0) is wider for irregularly shaped blocks of rock than for the more uniform concrete spheres and cubes.

These results indicate that the relationship given by Equation (3.17) can be used to estimate values for the angular velocity of blocks with known dimensions and velocities.

Figure 3.10 shows that for actual rock falls, considerable scatter occurs in the value of ω, and it is likely that ω will sometimes increase and sometimes decrease during impact. This variation in the angular velocity depends on the attitude of the irregular block as it impacts the rough slope surface. Figure 3.11 shows two successive impacts (n, $n + 1$) of an irregular block, with frictional resistance R, generated at the impact points. At point n, the normal and tangential velocity components generate moments about the impact points that tend to increase the angular velocity, while at point ($n + 1$), the moments produced by the two velocity components are in opposite directions and tend to decrease ω.

Further discussion on rotational velocity is provided in Section 4.5 related to changes in rotational velocity during impact produced by friction acting at the contact during impact.

3.4.2 Relationship between Trajectories and Angular Velocity

For a perfectly elastic impact of a smooth, nonrotating block, the impact and restitution velocities and angles will be identical. However, for actual rock falls where the block is rotating with frictional resistance being developed at the impact point and plastic compression of the slope materials occurring, the restitution parameters will change during impact.

Impact mechanics theory discussed in Chapter 4 shows that the effect of clockwise rotation as shown on Figure 3.11, for a frictional impact is to flatten the trajectory, i.e., reduce the value of θ_0. Also, the higher the clockwise angular velocity, the smaller the value of θ_0 with a flatter trajectory. Furthermore, as the trajectory becomes flatter (closer to the slope surface), the velocity increases as shown in Figure 3.8 where two trajectory paths are compared.

Impact mechanics also shows that the effect of counterclockwise rotation is to produce higher, slower trajectories.

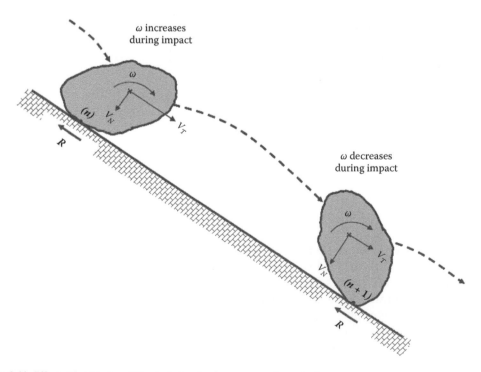

Figure 3.11 Effect of attitude of block during impact on angular velocity.

3.5 FIELD OBSERVATIONS OF ROCK FALL TRAJECTORIES

This section briefly describes some of the characteristics of rock fall trajectories in the field.

3.5.1 Rock Falls down Gullies

Rock falls on steep slopes are similar to water flow in that falls will tend to concentrate in gullies. That is, minor gullies in the upper part of the slope act as a "watershed" in which falls over a wide area on the upper slope combine into a single gully at the base of the slope. For example, at one active rock fall site, falls were originating from a maximum height of about 250 m (820 ft) over a slope length of about 120 m (400 ft), but they were almost all contained by a 12 m (40 ft) long fence in a gully at the base of the slope. This shows that careful examination of the rock fall geometry can result in substantial savings in construction of protection measures.

Another effect of rock falls concentrating in gullies is that the fall path is not a straight line down the maximum gradient line. Figure 3.12 shows a slope about 500 m (1,640 ft) high on which rock falls are entirely concentrated into three sinuously shaped gullies in which the rock fall paths are significantly longer and flatter than the slope cross section. This shows that rock falls can be much more accurately modelled in three dimensions than in two dimensions. That is, cross sections of the slope shown in Figure 3.12 would be irregular where the section crossed the gullies, and modeling of this slope would probably show significant trajectory heights where falls impacted the sides of the gullies. In contrast, a section down a gully would be an essentially uniform slope in which trajectory heights would probably not exceed 2 m (6.6 ft).

Figure 3.12 also shows how all rock falls over a length of several hundred meters along crest can be contained by just three fences located in the base of the gullies, each about 20 m (65 ft) long.

Figure 3.12 Mountain slope with three sinuous gullies in which all rock falls are concentrated.

3.5.2 Run-Out Distance

The maximum distance that a rock fall will travel from the source zone can be an important parameter in the location and/or protection of facilities in the run-out area. Figure 1.1 shows a typical rock fall site made up of four areas—the rock face where the fall originates, a colluvium slope, a talus slope where most of the rock falls accumulate, and a run-out zone, or rock fall "shadow area," between the base of the talus cone and the maximum travel distance. The maximum travel distance is defined by a line drawn at a dip angle of 27.5 degrees from the base of the rock face to the intersection with the ground surface (Hungr and Evans, 1988).

Depending on the level of risk acceptance for the facilities, the run-out area could be designated an exclusion zone in which no continuously occupied structure such as a house, could be located. Alternatively, it may be acceptable to locate such facilities as a lightly used road or a golf course within the run-out zone, perhaps with some protection such as a ditch along the up-slope edge.

Section 8.5.5 discusses the application of decision analysis to rationally evaluate the cost-benefit of alternate rock fall protection measures based on the probability of rock falls occurring, the consequence of such falls, and the cost of constructing protection measures.

3.5.3 Dispersion in Run-Out Area

Where a talus cone has developed at the base of a rock fall area, over time falls will disperse uniformly over the talus to build up the cone equally over its full area. This is a progressive process whereby the accumulation of rock falls on one area of the cone will then divert subsequent rock falls to a lower area that is built up in turn.

At the Ehime test site discussed in Chapter 2, it was found that the falls dispersed over an area subtended by an angle of 60°, with about 93% of the falls within a 30° angle. At other test sites in Japan, the angle defining the width of the dispersal area varied from 45° to 70° (Ushiro and Hideki, 2001). At Tornado Mountain, the horizontal distance between the two rock falls was 87 m (285 ft) after falling 740 m (2,450 ft), or a dispersion angle of 7°.

This information can be used to determine the length of rock fall fence that may be required to contain falls.

Chapter 4

Impact Mechanics

The theory of impact mechanics is used in a wide variety of fields (Goldsmith, 1960; Stronge, 2000) and builds on earlier work carried out by Sir Isaac Newton (Newton, 1687) and others such as Poisson and Hertz in the 19th century. The theories generally apply to the impact between two bodies, made of different materials, that are both translating and rotating, have unequal masses, and are moving in three-dimensional space. For rock falls, the impact conditions are somewhat simplified because one of the bodies (the slope) is stationary and has infinite mass. However, the roughness of the slope and irregularity of the rock falls introduce complexities in the modeling that can be accounted for by probabilistic analyses. Furthermore, the theory needs to account for the condition that the two bodies may not be of the same material.

This chapter summarizes the application of impact mechanics theory to rock falls, and how the theory can be applied to the actual rock fall events documented in Chapter 2. Chapters 5 and 6 cover, respectively, coefficients of restitution and energy losses, and show how the field results can be used to calibrate impact mechanics theory and modeling programs.

The theory for impact mechanics described in this book is based primarily on the work of W. J. Stronge (2000).

4.1 PRINCIPLES OF RIGID-BODY IMPACT

The theory of impact mechanics can be applied to rock falls in order to understand the impact process, and to develop algorithms for modeling rock falls. This work involves the application of rigid-body impact and kinetics to rock fall behavior as described below.

4.1.1 Rigid-Body Impact

The particular physical conditions that are applicable to rock falls are low velocity (i.e., less than about 40 m · s⁻¹ (130 ft · s⁻¹)), and impact of initially nonconforming hard, rigid bodies that result in minor deformations but high stresses generated over the small area of the impact, and to a very shallow depth. During impact, the shapes of the two surfaces are briefly conformable. These contact conditions, in which no interpenetration or adhesion of the two bodies occurs, are referred to as *low compliance* impacts.

The highly stressed contact area of the rock fall and the slope can be considered as a short, stiff spring, or an infinitesimally small, deformable particle (Figure 4.1). The spring or particle is compressed during the compression phase of impact and then releases energy to force the rock fall away from the slope after the time of maximum compression and during the restitution phase of impact.

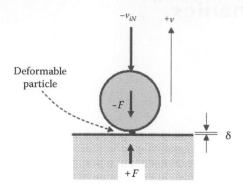

Figure 4.1 Forces generated at contact point during normal impact.

The duration of the impact is very short, possibly a few tenths of a second, which has a number of implications for modeling rock falls. First, changes to the position of the rock fall are negligible during impact, and, second, gravitational forces can be ignored because they are very low compared to the high induced forces at the impact point.

Based on these assumed conditions of impact, the change in velocity of the rock fall during impact can be resolved as a function of the normal component of impulse, where the normal impulse is equal to the integral of the normal contact force over the time of the impact. Since the impact involves only compression, and not extension of the spring or particle, the normal component of impulse is a monotonously increasing function of time after impact. Thus, variations in velocity during impact are resolved by choosing as an independent variable the normal component of impulse, p_N, rather than time. This principle gives velocity changes during impact that are a continuous, smooth function of impulse (Stronge, 2000).

Throughout this book, impacts are modeled in terms of the relationship between changes in normal and tangential relative (between the periphery of the body and the slope) velocity components, v_N and v_T, and the normal component of impulse, p_N.

4.1.2 Kinetics of Rigid Bodies

Kinetics is a means of examining the change in velocity of the body when forces act on the body during impact.

A rock fall is a rigid body as defined in Section 4.1.1 above and can be considered as a point mass, of infinitesimal size. If the body has mass m and is moving with velocity V (at the center of gravity), then the momentum of the body is $(m \cdot V)$. If a resultant force F acts on the body, this causes a change in momentum according to Newton's second law of motion.

> *Second Law*: the momentum $(m \cdot v)$, of a body has a rate of change with respect to time that is proportional to, and in the direction of, any resultant force $F(t)$ acting on the body.

Assuming that the mass of the body is constant during impact, then the change in velocity is a continuous function of the impulse, p. The forces F, $-F$ acting on the rock fall and the slope that prevent interpenetration of the two bodies are related by Newton's third law of motion:

> *Third law*: two interacting bodies have forces of action and reaction that are equal in magnitude, opposite in direction and collinear, that is, $F = -F$.

The application of the second and third laws to rock falls is shown in the following sections.

4.2 FORCES AND IMPULSES GENERATED DURING COLLINEAR IMPACT

Calculation of the forces and impacts during rock fall impacts applies Newton's second and third laws discussed in Section 4.1.2 above.

The impact process for a nonrotating, rigid body moving with relative velocity v_N and impacting a stationary surface at right angles can be simulated with an infinitesimal, deformable particle at the contact point between the two bodies (Figure 4.1). The particle acts as a short, stiff spring that, during impact, generates equal and opposite reaction forces F, $-F$ at the point of impact that are parallel to the velocity direction.

The reaction forces change the momentum of the body, and for a constant mass during impact, the velocity is changed. The change in momentum generated by the impact produces a finite impulse p that continuously changes the velocity during the impact time. During impact of a body with mass m, a change in normal velocity occurs from v_{iN} at impact (time $t = i$) to v_N at time t. The normal component of impulse p_N, generated by the normal component of the contact force $F(t)$ is given by the relationship:

$$dp_N = m(v_N - v_{iN}) = F\,dt \tag{4.1a}$$

where

$$F = m\frac{dv}{dt} \tag{4.1b}$$

$$\therefore \quad dp_N = m\frac{dv}{dt}dt$$

$$\therefore \quad dv_N = \frac{dp_N}{m}$$

and

$$m(v_N - v_{iN}) = \int_i^t F\,dt = p_N \tag{4.1c}$$

The relative normal velocity v_N at any time t during the impact can be obtained by integration, for the limit that at the moment of impact $t(i)$, the normal velocity is given by:

$$v_N = \int_{t(i)}^t \frac{1}{m}dp_N \tag{4.2}$$

$$v_N = v_{iN} + \frac{p_N}{m} \quad \text{where } v_{iN} < 0 \tag{4.3}$$

The impacting normal velocity is negative because, as shown in Figure 4.1, the positive normal axis is in the direction away from the point of impact.

Equation (4.3) shows that the normal component of the relative velocity is a linear function of the normal impulse. This relationship between v and p is the basis for determining changes in impulse during impact, and finding the terminal impulse p_f, at the termination of impact $(t = f)$ when the body separates from the slope.

Equation (4.3) can also be used to find the impulse at maximum compression (p_{cN}). At the point of maximum compression, the normal velocity is momentarily equal to zero and the corresponding normal impulse has a value p_{cN} given by Equation (4.3)

$$0 = v_{iN} + \frac{p_{cN}}{m}$$

and

$$p_{fN} = -m \cdot v_{iN} \tag{4.4}$$

At the end of the impact $(t = f)$, the final normal velocity is v_{fN} and the final normal impulse (p_{fN}) can also be found from Equation (4.3):

$$p_{fN} = (m \cdot v_{fN} - m \cdot v_{iN}) \tag{4.5}$$

4.3 ENERGY CHANGES DURING IMPACT

The impact process results in compression δ of the deformable particle during the compression phase (Figure 4.1), followed by expansion during the restitution phase. The changes in the normal contact force F during impact are illustrated in Figure 4.2(a) where the force (F_c) and deformation (δ_c) are at a maximum at the end of the compression phase, followed by partial recovery (δ_f), for inelastic impact, at the completion of the recovery phase. The recovery of kinetic energy is the process that drives the bodies apart in the final phase of the impact after maximum compression.

Figure 4.2(b) shows the changes in the normal contact force as a function of time. The area under the $[t - F]$ curve up to time t_c is the impulse p_c generated during the compression phase and represents the kinetic energy of relative motion that is converted into internal energy of deformation. The area between times t_c and t_f the change in impulse $(p_f - p_c)$ and represents the energy recovered during the restitution phase. The changes in velocity during impact can be quantified in terms of the normal coefficient of restitution, e_N, that is, the ratio of the final normal velocity v_{fN} to the impact normal velocity, v_{iN}.

On Figure 4.2(b), for an elastic impact the two areas are identical—$e_N = 1$, while for a perfectly plastic impact no energy is recovered—$e_N = 0$.

A fully plastic impact in which no recovery of energy occurs is shown in Figure 4.3 where a rock fall is embedded in a wall constructed with gabions $(e_N = 0)$. In this case, almost all the impact energy has been absorbed by the plastic deformation of the gabions, with only a little energy being absorbed by the elastic deformation of the rock fall. The design of MSE (mechanically stabilized earth) rock fall barriers is discussed in Section 10.2.

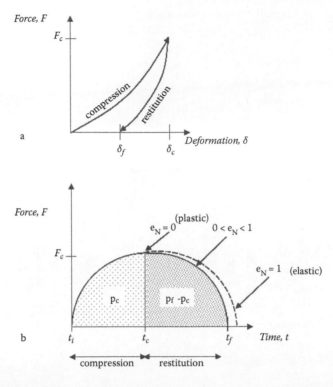

Figure 4.2 Variation in force F during impact. (a) Relationship between force and deformation at the impact point; (b) change in force and impulse with time during impact; p_c is impulse generated up to time of maximum compression ($t = i$ to $t = c$); $p_f - p_c$) is impulse generated during restitution phase of impact ($t = c$ to $t = p_c$).

4.4 COEFFICIENT OF RESTITUTION

The principle of separating the compression and restitution phases of impact can be demonstrated on a normal impulse [(p_N)—relative velocity (v)] plot as shown in Figure 4.4. On this plot, the normal velocity changes during impact, starting with a negative value ($-v_{iN}$) at the point of impact, increasing to zero at the point of maximum compression p_c, and finally reaching a positive value (v_{fN}) at the point of separation. Also, the tangential velocity v_T decreases continuously during impact from v_{iT} at the point of impact, to v_{fT} at the point of separation, as the result of frictional resistance on the contact surface. The role of friction on impact behavior is discussed in Section 4.5.

The [$p_N - v$] plot on Figure 4.4 shows the changes in both the normal (v_N) and tangential (v_T) velocities, and the magnitude of the internal energy of deformation generated during impact. These changes in velocity and energy can be quantified in terms of the coefficient of restitution,[*] e that has normal and tangential components as follows:

$$e_N = -\frac{v_{fN}}{v_{iN}} \tag{4.6}$$

---

[*] In this treatment of impact mechanics as it applies specifically to rock falls, the term *coefficient of restitution* is used to quantify the changes in both velocity and energy, and it encompasses the terms kinetic, kinematic, and energetic coefficients of restitution that apply in certain impact conditions (Stronge, 2000).

Figure 4.3 Example of fully plastic impact where a rock fall is embedded in a gabion wall.

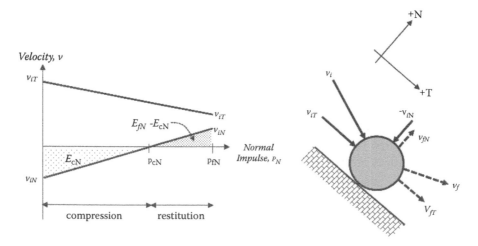

Figure 4.4 Relationship between normal impulse p_N and changes in tangential and normal velocities v_T, v_N, and energy during impact; E_{cN} is the kinetic enrgy absorbed during the compression phase of impact ($t = c$); ($E_{fN} - E_{cN}$) is the strain energy recovered during the restitution phase ($t = c$ to $t = f$).

and

$$e_T = \frac{v_{fT}}{v_{iT}} \tag{4.7}$$

The parameter e_N, which is related to compression/hystersis at the contact point, is used to determine the normal velocity and energy changes that occur during impact as discussed in this chapter. The parameter e_T, which quantifies the changes in tangential velocity during impact, is a function of the friction acting on the contact surface and the effect of friction on the angular velocity of the body.

Examples of actual changes in normal and tangential velocities during impact and the corresponding values of e_N and e_T are discussed in the documentation of rock fall events discussed in Chapter 2, and are listed in Table 2.1.

With respect to the energy of deformation, the triangular area E_{cN} in Figure 4.4 represents the kinetic energy of normal motion that is absorbed in compressing the deformable region, while triangular area $(E_{fN} - E_{cN})$ represents the elastic strain energy recovered during restitution. The expression for the energy change during compression is:

$$E_{cN} = \int_0^{p_{cN}} v \, dp = \int_0^{p_{cN}} \left(v_{iN} + \frac{p_N}{m} \right) dp$$

$$= \left(v_{iN} \cdot p_{cN} + \frac{p_{cN}^2}{2m} \right)$$

since $p_{cN} = -m \cdot v_{iN}$ (Equation [4.4])

$$E_{cN} = -\frac{1}{2} m \cdot v_{iN}^2 \tag{4.8}$$

and the energy change during restitution is:

$$(E_{fN} - E_{cN}) = \int_{p_{cN}}^{p_{fN}} \left(v_{iN} + \frac{p_N}{m} \right) dp = \left[v_{iN} \cdot p_N + \frac{p_N^2}{2m} \right]_{p_{cN}}^{p_{fN}} \tag{4.9}$$

$$= \frac{1}{2} m \cdot v_{iN}^2 \left(\frac{p_{fN}}{p_{cN}} - 1 \right)^2$$

where $v_{iN} < 0$.

Derivation of Equations (4.8) and (4.9) is shown in Appendix I. Equations (4.8) and (4.9) are also derived in Chapter 6 that discusses energy losses during impact.

The expressions in Equations (4.8) and (4.9) for the partially irreversible changes in kinetic energy of normal motion that occur during impact can be used to define the normal coefficient of restitution, e_N, as follows:

$$e_N^2 = -\frac{E_N(p_{fN}) - E_N(p_{cN})}{E_N(p_{cN})} \tag{4.10}$$

This definition of the coefficient of restitution where the impulse is an independent variable separates the energy loss due to compression and hysteresis of the contact forces from that due to friction and slip between the colliding bodies. As shown in Figure 4.2, the value for e_N can range from $e_N = 1$ for a perfectly elastic material where no loss of energy occurs during impact, to $e_N = 0$ for a perfectly plastic material where no separation occurs during impact, and no recovery of the initial kinetic energy occurs.

Also, the relationships shown in Equations (4.8), (4.9), and (4.10) can be combined, for normal impact, to find the following expression for the normal coefficient of restitution in terms of the normal impulses at maximum compression (p_{cN}) and at the completion of the impact (p_{fN}):

$$e_N^2 = \left(\frac{p_{fN}}{p_{cN}} - 1 \right)^2 \tag{4.11}$$

Substitution of the expressions for p_{cN} and p_{fN} in Equations (4.4) and (4.5) into Equation (4.11), yields the following expression relating impulse to the coefficient of restitution:

$$p_{fN} = -m \cdot v_{iN}(1 + e_N) = p_{cN}(1 + e_N) \tag{4.12}$$

and for the normal coefficient of restitution,

$$e_N = -\frac{v_{fN}}{v_{iN}} \tag{4.6}$$

$$e_N = \frac{(p_{fN} - p_{cN})}{p_{cN}} \tag{4.13}$$

As shown in Figure 4.4 and expressed in Equations (4.6) and (4.13), the normal coefficient of restitution is the ratio of final normal velocity to the impact normal velocity, and is the square root of the negative ratio of the energy recovered during the restitution phase of the impact to the energy lost during the compression phase (Equation (4.10)).

For rough bodies where slip occurs at the contact point, but the direction of slip is constant, the two expressions for the coefficient of restitution in Equations (4.6) and (4.13) are equivalent.

4.5 FRICTIONAL ANGULAR VELOCITY CHANGES DURING IMPACT FOR ROUGH SURFACE

For oblique impact of a rock fall with a slope, the impulse acting at the contact point gives an impulsive moment about the center of mass of the rock fall (Figure 4.5). This impulsive moment changes the angular and translational velocities during the period of contact.

Changes in the rotational and translational velocities during impact can be attributed to friction at the rock-slope contact. According to Coulomb's definition of friction, the coefficient of friction μ, is the ratio of the tangential to normal forces acting at the contact point. Furthermore, the ratio of the tangential to normal forces is a constant such that the friction coefficient is also a constant that is independent of the slip velocity and the normal force.

It is noted that in rock mechanics, it is usually assumed that the coefficient of friction decreases with increasing normal stress due to break down of the asperities on the rough rock surfaces during shear movement. However, for rock fall impacts that are of short duration and have limited shearing distance, the friction coefficient is considered to be independent of the normal force.

When a translating and rotating body impacts a stationary slope, slip will initially occur between the two bodies, and frictional forces will be generated at the contact point. In accordance with the usual concept of friction relating normal and tangential impulses, slip of the body on the slope is assumed to occur at the value of limiting friction, requiring that

$$p_T = |\mu \cdot p_N| \tag{4.14a}$$

while slip stops and the body rolls during impact when

$$(p_T \leq \mu \cdot p_N) \tag{4.14b}$$

This transition from slip to rolling is illustrated on Figure 4.5. The behavior of a rotating body during impact can be demonstrated on the [normal impulse (p_N)–relative velocity (v)] plot on Figure 4.5. For a spherical body with radius r and rotating at angular velocity $-\omega$, the velocity at the periphery of the body $(-r \cdot \omega)$ is parallel, but opposite in direction, to the translational tangential velocity v_T. If the magnitudes of $|r \cdot \omega|$ and v_T are unequal, then slip will occur between the moving body and the stationary ground, with the slip velocity v_S, being given by:

$$v_S = v_T + r \cdot \omega \tag{4.15}$$

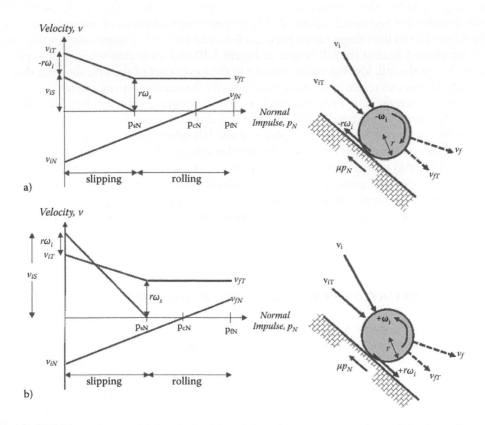

Figure 4.5 Changes in rotational (ω) and slip (v_s) velocities during impact, and transition from slip to rolling mode when $v_s = 0$. (a) Negative angular velocity: $v_S = (v_T - r \cdot \omega)$; (b) positive angular velocity: $v_S = (v_T + r \cdot \omega)$.

The direction of rotation is negative if the peripheral velocity $(r \cdot \omega)$ is in the opposite direction to the positive tangential axis. For rock falls, the angular velocity is usually negative because the tangential translational velocity acting downslope and the friction force $(\mu \cdot p_N)$ acting upslope interact to spin the rock clockwise as shown in Figure 4.5(a).

The slip velocity depends on the both the magnitude and direction of the angular velocity relative to the tangential velocity as defined by Equation (4.15) and shown on Figure 4.5. If the angular velocity is negative, then the slip velocity is less than the initial tangential velocity (Figure 4.5(a)), while if the angular velocity is positive, the slip velocity is greater than the initial tangential velocity (Figure 4.5(b)).

Depending on the details of the impact process, the values of $(r \cdot \omega)$ and v_T may equalize during impact, in which case slip will cease at impulse p_{sN} and from this point the body will roll with no change in angular velocity. In other conditions related to the attitude of the irregular body impacting the slope, slip may continue throughout the impact.

The $[p_N - v]$ plot in Figure 4.5 also shows that the tangential velocity decreases during the slip phase of the impact from v_{iT} to v_{fT}. This reduction in tangential velocity can be attributed to the friction force acting at the contact between the impacting body and the slope, and to the irregularities of the slope at the impact point. Equations developed in Section 4.6 show the slope of the tangential velocity line.

The $[p_N - v]$ plots in Figure 4.5 show the angular velocity increasing during the slip phase $(\omega_s > \omega_i)$. However, for irregularly shaped, rotating blocks impacting the slope at a variety of attitudes, the velocity components may combine under some conditions to reduce the velocity during impact, rather than increasing with each impact (see also Figure 3.11). At the Ehime test site described in Sections 2.1.3 and 2.2.1, the rotational velocities of about 100 test blocks on the 42 m (140 ft) high slope varied between 6 and 33 rad · s^{-1}. The measured values of ω at Ehime are plotted against the fall height on Figure 3.10 and show that while ω does generally increase during the fall, low values for ω occur in the lower part of the fall, showing that some impact conditions can result in the rotational velocity decreasing during impact.

Another characteristic of a rotating body is that the angular velocity changes only during impact as the result of the frictional force that acts during contact between the body and the slope. However, during the trajectory phase of the fall when contact between the body and the slope ceases, no forces act to change the angular velocity that remains constant during the trajectory. This also means that the rotational energy remains constant during the trajectory.

The effect of friction during contact can be understood by considering that for a completely frictionless, smooth contact, no change in tangential or angular velocity occurs during impact because no shear resistance is generated at the contact surface. That is, the v_T and v_S lines are horizontal on the $[p_N - v]$ plot for frictionless contact. However, v_N will still change during impact because of energy losses produced by compression of the body and slope at the contact point.

4.6 IMPACT BEHAVIOR FOR ROUGH, ROTATING BODY

When a rock fall impacts a slope, the parameters defining the impact conditions are the translational and rotational velocities at the moment of impact $(t = i)$, and the impact angle relative to the slope (θ_i). In addition, the characteristics of the slope are defined by the normal coefficient of restitution, and the friction coefficient. In order to model rock fall behavior, it is necessary to calculate the final translational and rotational velocities $(t = f)$, and the angle at which the body leaves the slope. These three final parameters can then be used to calculate the subsequent trajectory of the body. This section describes how impact mechanics can be used to derive the equations defining changes in velocity components and angles during impact. The impact mechanics model discussed in this section is based on the work of Stronge (2000) and Goldsmith (1960) on collision between solid bodies, with modifications to suit the particular conditions of rock falls.

The theory of impact mechanics can be used to model impact between two irregularly shaped, rough, rotating bodies moving in three-dimensional space; for rock falls, the model can be simplified by making one of the bodies (the slope) stationary and of infinite mass. For the purpose of developing equations to model impact in this book, further simplifications will be made that friction is developed at the impact point, and the body is moving in two-dimensional space (plane motion; Figure 4.6). The size and shape of the body are defined by the radius (r) and the radius of gyration (k).

The equations of motion are referenced to coordinate axes set up at the impact point in directions normal N and tangential T to the slope surface. The positive normal axis is away from the slope and the positive tangential axis is downslope such that the impact normal velocity is negative and the tangential velocity is positive. In addition, the angular velocity is negative when it is in the $-T$ direction at the contact point. Subscript "i" refers to conditions at impact (time, $t = i$), and subscript "f" refers to restitution conditions at the end of impact (time, $t = f$). The radius of the spherical body is r.

4.6.1 Impulse Calculations

The impact of the rock fall with the slope in plane motion produces normal and tangential impulses p_N and p_T, respectively, at the contact point that alter the normal and tangential velocity components of the rock during impact. The velocity changes are governed by the magnitude of the normal coefficient of restitution e_N and the limiting value of the coefficient of friction μ. Values of the coefficient of restitution have been determined by documenting actual rock fall events as described in Chapter 2, and for the coefficient of friction from both laboratory tests and rock fall events.

Figure 4.6 shows a spherical body with mass m, radius r, and radius of gyration k, impacting the slope at a shallow angle. Following the nomenclature defining impact velocities shown in previous sections of this chapter, at any time during the impact, the linear impulse-momentum Equation (4.1c) provides the following relationship for the normal impulse p_N related to the change in the normal component of the velocity:

$$m(v_N - v_{iN}) = p_N \tag{4.1c}$$

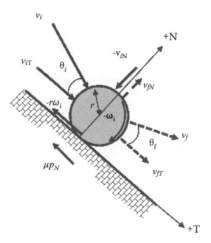

Figure 4.6 Impact of rough, rotating sphere on a slope in plane motion.

And the equivalent tangential impulse equation p_T :

$$m(v_T - v_{iT}) = -p_T \tag{4.16}$$

where v_{iT} and v_{iN} are, respectively, the initial tangential and normal translational velocity components at impact (time, $t = i$).

With respect to the rotation of the block with angular velocity ω at time t, initial angular velocity ω_i and moment of inertia I, the angular impulse-momentum equation is

$$I(d\omega) = I(\omega - \omega_i) = \int_0^t Fr \, dt \tag{4.17}$$

and

$$m \cdot k^2 (\omega - \omega_i) = p_T \cdot r \tag{4.18}$$

where $I = m \cdot k^2$, and k is the radius of gyration of the body.

The volumes and radii of gyration of bodies that may simulate rock falls are listed in Table 4.1.

4.6.2 Final Velocities for Rock Fall Impacts

For a rock fall impacting a slope at an oblique angle, the rock will have both translational v, and angular ω, velocities, with the translational velocity expressed as normal and tangential components relative to the slope surface (Figure 4.6). At the contact point, equal and opposite forces, F, $-F$ are developed that oppose interpenetration of the rock into the slope and give differentials of impulse dp in the normal and tangential directions that are related by:

$$dp = F \, dt \tag{4.1a}$$

Table 4.1 Volume and radius of gyration of common rock fall body shapes

Body shape	Volume, Ω	Axial radius of gyration, k
Cube: side length L	L^3	$\dfrac{L}{\sqrt{6}}$ —axis normal to face, or along diagonal
Sphere: radius r	$\dfrac{4}{3}\pi r^3$	$\left(\dfrac{2}{5}\right)^{0.5} r$
Cylinder: length L, radius r	$\pi r^2 L$	$\left(\dfrac{3r^2 + L^2}{12}\right)^{0.5}$ —axis through mid-height ("tumbling" motion)
Ellipsoid: axes $2a$, $2b$, $2c$	$\dfrac{4}{3}\pi abc$	$\left(\dfrac{(a^2 + b^2)}{5}\right)^{0.5}$ —rotation about axis c

Newton's second law (see Section 4.1.2), gives equations of motion for translation of the centre of the rock fall mass in the normal N, and tangential T, planes:

$$dV_N = \frac{dp_N}{m}$$

and

$$dV_T = \frac{dp_T}{m}$$

and for planar rotation of the rock fall:

$$d\omega = \frac{r}{m \cdot k^2} dp \qquad (4.19)$$

The impact mechanics principles discussed in this chapter of relating velocity and impulse changes during impact can be used to derive equations for the final tangential and normal velocity components. Appendix II shows the method of deriving the equations, for the case of frictional impact where transition from sliding to rolling occurs during impact. For these conditions, the expressions for the final velocity components at the centre of mass for a spherical body are as follows:

$$V_{fT} = V_{iT} - \frac{(V_{iT} + r \cdot \omega_i)}{(1 + r^2/k^2)} \qquad (4.20)$$

$$V_{fN} = -V_{iN} \cdot e_N \qquad (4.21)$$

and the final rotational velocity is:

$$\omega_f = \omega_i - \frac{r}{k^2} \frac{(V_{iT} + r \cdot \omega_i)}{(1 + r^2/k^2)} \qquad (4.22)$$

Equations (4.20) and (4.21) for the final tangential and normal velocity components, respectively, can then be solved to find the final restitution velocity v_f and angle θ_f as follows:

$$V_f = \sqrt{V_{fT}^2 + V_{fN}^2} \qquad (4.23)$$

$$\theta_f = atan\left(\frac{V_{fN}}{V_{fT}}\right) \qquad (4.24)$$

Figure 4.7 shows the final velocities and angles diagrammatically, in terms of the three impact parameters v_i, θ_i, and ω_i, and the size and shape of the body (r, k).

It is of interest that in Equations (4.20) to (4.24) defining the final velocities and restitution angle, the only physical property of the slope/rock fall that is incorporated is the normal coefficient of restitution. The friction coefficient, which relates tangential and normal impulses during slip (Equation (4.14[a]) is used to calculate the reduction in tangential velocity during impact as shown in Appendix II, Section II.2. Also, Chapter 6 shows the calculation of energy losses for rotating bodies incorporating the coefficient of friction.

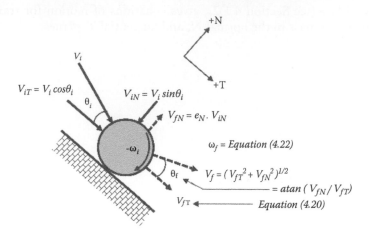

Figure 4.7 Diagram of impact showing equations defining impact and restitution velocity vectors.

When slip halts and the block rolls, no additional change in tangential velocity occurs because sliding friction does not operate in this phase of the impact process. Furthermore, while the size (radius, *r*) and shape (radius of gyration, *k*) of the body influences the final velocity and angle calculations, these values are independent of the mass because the forces generated at the impact point are much larger than gravity so body forces can be neglected (see Section 4.1). The equations presented in this chapter demonstrate the basic behavior of a rough, rotating sphere impacting a stationary slope surface, for a two-dimensional (planar) condition. If necessary, the equations can be modified to analyze other block shapes such as ellipsoids or slabs. While the equations are a simplification of actual rock fall behavior, they provide a useful framework for understanding the various factors that influence rock fall behavior, and can be used to examine actual rock falls and assist in the design of protection measures. Section 4.7 compares actual velocities documented in Chapter 2 with velocities calculated from the theoretical equations.

4.6.3 Example of Impact Mechanics Calculation

Correct application of the equations in this chapter requires careful attention to the signs (negative and positive) of the velocities and angles according to the system of axes as shown in Figure 4.6. The following worked example illustrates a sample calculation.

Worked Example 4A—impact final velocities: for a 1.5-m-diameter spherical rock fall, the impact translational and rotational velocities (V_i and ω_i), and impact angle θ_i relative to the slope, are shown in Figure 4.8. The tangential and normal components of the impact velocity are calculated as follows:

$$V_{iT} = V_i \cos\theta_i = 19.9\, ms^{-1}; \; V_{iN} = V_i \sin\theta_i = -9.3\, ms^{-1}$$

The radius of gyration *k* of the sphere is $r\sqrt{(2/5)} = 0.47$ m.

The coefficient of restitution e_N is determined from the relationship between θ_i and e_N as discussed in Chapter 5 (Section 5.2, Equation [5.4]) where it is demonstrated that

$$e_N = 19.5\theta_i^{-1.03}$$

$$= 0.71$$

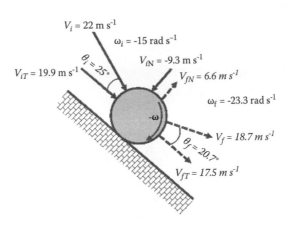

$V_i = 22$ m s^{-1}

$\omega_i = -15$ rad s^{-1}

$V_{iN} = -9.3$ m s^{-1}

$V_{iT} = 19.9$ m s^{-1}

$\theta_i = 25°$

$V_{fN} = 6.6$ m s^{-1}

$\omega_f = -23.3$ rad s^{-1}

$-\omega$

$V_f = 18.7$ m s^{-1}

$\theta_f = 20.7°$

$V_{fT} = 17.5$ m s^{-1}

Figure 4.8 Example of rock fall impact showing values calculated final (restitution) velocity and angle.

Application of Equations (4.20) to (4.24) gives the values for the final translational and angular velocity components, and the restitution as shown in Figure 4.8. These results show that during impact the normal velocity component changes from −9.3 m · s^{-1} toward the slope to +6.6 m · s^{-1} away from the slope, while the tangential velocity component reduces from 19.9 to 17.5 m · s^{-1}, and the angular velocity increases, in the negative direction, from −15 to −23.3 rad · s^{-1}. The overall effect of the impact is to reduce the velocity of the rock by 3.3 m · s^{-1}, or 15%. The final angle, $\theta_f = 20.7$ degrees.

The effect of rotation on the impact process is illustrated by letting $\omega_i = 0$, for which the new values for the calculated final velocities are: $V_{fT} = 14.2$ m · s^{-1}, $\omega_f = -19$ rad · s^{-1} and $\theta_f = 24.8$ degrees. That is, the effect of a rotational velocity of −15 rad · s^{-1}, compared to the body not rotating at impact, is to increase V_{fT} from 14.2 to 17.5 m · s^{-1}, and produce a flatter trajectory as θ_f changes from 24.8 to 20.7 degrees.

4.6.4 Effect of Angular Velocity on Trajectories

Worked example 4A and Figure 4.8 show the influence of the angular velocity on the restitution velocity and angle. That is, an increasing negative angular velocity produces a flatter, greater velocity trajectory. The impact mechanics equations can also be used to find the effect of positive rotational velocity on trajectories as follows.

Equation (4.20) shows the relationship between the impact rotational velocity ω_i and the restitution tangential velocity V_{fT}. Also, Equation (4.21) relates the restitution normal velocity v_{fN} to the normal coefficient of restitution e_N, and Equations (4.23) and (4.24) can then be used to find the restitution velocity V_f and angle θ_f.

The influence of the impact angular velocity ω_i on the restitution velocity and angle is shown in Figure 4.9. In these three models, the impact velocity is $V_i = 22$ m · s^{-1} at angle $\theta_i = 25°$. Impacts have been studied for three angular velocities: Model (a) – $\omega_i = -15$ rad · s^{-1}; Model (b) – $\omega_i = -25$ rad · s^{-1}; Model (c) – $\omega_i = +15$ rad · s^{-1}. As shown on Figure 4.5, the angular velocity is negative if its direction at the contact point is in the opposite direction to the positive tangential axis, that is, clockwise in this model.

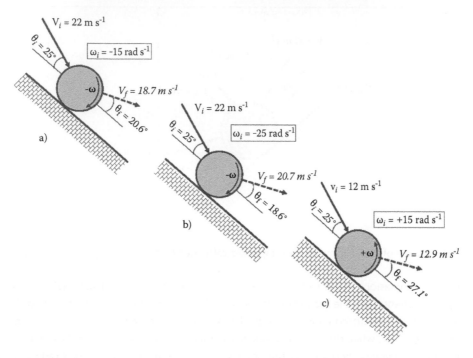

Figure 4.9 Influence of impact angular velocity, ω_i on restitution velocity, v_f and angle, θ_f. (a) $\omega_i = -15$ rad · s⁻¹; (b) $\omega_i = -25$ rad · s⁻¹; (c) $\omega_i = +15$ rad · s⁻¹.

The calculation of the final velocities and angles for the three conditions on Figure 4.9 confirms that the effect of an increasing negative angular velocity is to increase the final velocity and produce a flatter trajectory. That is, for conditions (a) and (b), the final velocity increases from 18.7 m · s⁻¹ to 20.7 m · s⁻¹, while the final angle decreases from 20.6° to 18.6°.

In Model 4.9(c), the body is rotating in a counter-clockwise (positive) direction, and the restitution velocity now decreases to 12.9 m · s⁻¹ at a larger angle of $\theta_f = 27.1$ degrees.

In rock falls, rotation is almost always in the negative direction because the frictional resistance at the contact between the slope surface and the moving body sets up a moment acting clockwise. The calculations show that the negative angular velocity flattens the trajectory, which is consistent with the low trajectory heights shown on Figure 3.5. Furthermore, the distributions of θ_f values plotted in Figures 3.9(a) and (b) show that the majority of θ_f values are less than about 30°, with few high angle trajectories, that is, θ_f values greater than 60° are rare. An exception to this condition is for very shallow angle impacts ($\theta_i < 15°$) where high angle final trajectories can occur (see Section 5.2).

Calculation of ω_f using Equation (4.22) shows that for a sphere ω_f will increase with every impact. In fact, field studies show that the angular velocity may increase or decrease on impact, depending on the attitude of the nonspherical block relative to the slope at the point of impact. Measured values of ω_f shown on Figure 3.10 illustrate the variation in ω_f values and that ω_f may increase or decrease during impact after the initial 5 to 10 m fall height (see Section 3.4.1).

The trajectories of rock falls are important in the selection of appropriate heights for protection structures such as fences and barriers. Both field data and impact mechanics theory show that low trajectories are more common than high trajectories, and this may be taken into account in fence design.

4.7 CALCULATED VERSUS ACTUAL RESTITUTION VELOCITIES

Chapter 2 describes five rock fall events where the impact and restitution velocities and angles were carefully documented; vectors of these velocities for a typical impact at each site are shown on the insets in Figures 2.2 to 2.10. Information is also available on the shape, dimensions, radius, and radius of gyration of a typical block at each site.

These actual restitution velocities can be compared with velocities calculated using Equations (4.20) to (4.24) in Section 4.6.2 based on impact mechanics theory. In applying these equations, values are required for the initial angular velocity ω_i, and the normal coefficient of restitution e_N, at the impact points. Values for these two parameters were obtained from the following relationships:

$$\omega_i = \frac{V_i}{r} \tag{3.17}$$

where V_i is the impact velocity and r is the radius of the body.
and

$$e_N = 19.5\,\theta_i^{-1.03} \tag{5.4}$$

where θ_i is the impact angle.

Figure 4.10 is a plot comparing the two sets of restitution velocities. The plot shows that the calculated velocities are in all cases greater than the actual velocities by values ranging from 20% to 50%. Sensitivity analyses of these calculations show that the calculated and actual values are closer when the angular velocity is reduced and the normal coefficient of restitution is increased. However, reasonable changes to these parameters are not sufficient to make the calculated velocities equal to the actual values.

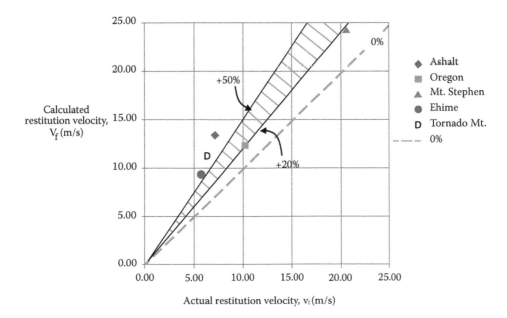

Figure 4.10 Plot comparing restitution velocities—actual (Chapter 2) and calculated (Equations (4.20) to (4.24)) values.

Chapter 5

Coefficient of Restitution

The results of the field studies in Chapter 2 and the discussion of impact mechanics in Chapter 4 illustrate how rock fall behavior can be defined by values for the normal and tangential coefficients of restitution (e_N and e_T) that are applicable to the particular site conditions. The coefficients of restitution quantify velocity changes during impact and help in understanding how conditions at the impact points influence rock fall behavior. Figure 5.1 shows three impacts and two trajectories and clearly illustrates the reduced velocity and height of the second trajectory due to the loss of energy at the first impact, that is, $e_N < 1$. Figure 5.1 shows the typical behavior of a rubber ball; rock falls will always have lower trajectories than these because of the relatively low e_N of rock.

Changes in the velocity components during impact and the corresponding coefficients of restitution are also illustrated in Figure 5.2 on the (normal impulse, p_N-relative velocity, v) diagram. The normal coefficient of restitution defines the change in normal velocity during impact and is related to the impact angle as discussed in this chapter and the inelastic compression characteristics of the slope materials. The tangential coefficient of restitution defines the reduction in tangential velocity during impact and is related to the friction force generated between the slope and the body. This chapter discusses how these coefficients are correlated with impact conditions and are not purely material properties.

This chapter summarizes the results of the e_N and e_T values obtained from the five field studies described in Chapter 2, encompassing four different slope materials and 57 impacts. A sixth location is a laboratory test where blocks of rock were dropped on to a concrete floor and the rebound heights were measured to give values for e_N.

5.1 NEWTON'S COEFFICIENT OF RESTITUTION

The concept of the coefficient of restitution, which in this case was the normal coefficient, was first developed by Isaac Newton (1686) who suspended spheres of the same material on pendulums and measured how high they rebounded after impact (Figure 5.3); the measurements incorporated corrections for velocity loses due to air friction. Values for the coefficient of restitution found by Newton included 0.56 for steel and 0.94 for glass. One of the purposes of the experiments was to prove the third law of motion: every action has an equal and opposite reaction.

It was assumed at the time of Newton's experiments that the coefficients of restitution were material properties. However, it is now understood that impact between rough, rotating bodies of different materials, such as rock falls, reductions in velocity depend not only on the material forming the body but also on the impact conditions such as mass and shape of the impacting body, and the impact angle and velocity.

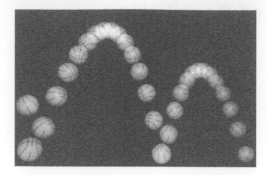

Figure 5.1 Impacts between successive trajectories showing typical inelastic behavior and loss of energy during impact where second trajectory (on right) is lower than first trajectory; rock falls will always have lower trajectories than those shown. (From Micheal Maggs, Wikimedia Commons.)

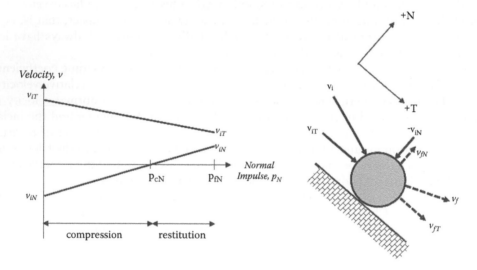

Figure 5.2 Normal impulse-relative velocity plot showing relationships between changes in normal (N) and tangential (T) velocity components and coefficients of restitution $-e_N = (v_{fn}/v_{iN})$ and $e_T = (v_{fT}/v_{iT})$.

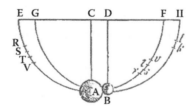

Figure 5.3 Isaac Newton's measurement of normal coefficient of restitution using impact of spheres suspended on pendulums.

5.2 NORMAL COEFFICIENT OF RESTITUTION

Equation (4.6) and the $[p_N - v]$ diagram shown in Figure 5.2 define the normal coefficient of restitution as the ratio of the final to impact normal velocities:

$$e_N = -\frac{v_{fN}}{v_{iN}} = \sqrt{\frac{h_f}{h_i}} \tag{4.6}$$

where v_{fN} $(t = f)$ and v_{iN} $(t = i)$ are the final (restitution) and impact velocities, respectively, and e_N is negative because v_{iN}, acting in the negative direction of the normal axis, has a negative value.

Equation (4.6) indicates that the value of the normal coefficient can be determined by dropping blocks of rock from height (h_i) on to different surfaces and measuring the height of the rebound (h_f) in a similar manner to the procedure used by Newton. However, this is not applicable to rock falls where the irregular, rotating blocks are impacting rough slope surfaces at oblique angles. In fact, for rock falls, it is difficult to measure the value of e_N in the laboratory because of the complexity of the impact process. Field measurement of coefficients of restitution involves carefully documenting impacts and trajectories by locating impact points or using a high-speed camera, and then calculating the normal and tangential components of the impact and restitution velocities.[*]

A further complication with the velocity measurements is that considerable scatter occurs in the values as the result of the wide variation in the impact conditions of the rotating, irregular blocks impacting the rough slope surface. This variation can best be handled using probabilistic methods in which the design values of velocity, coefficient of restitution, and rock fall mass are expressed as probability distributions rather than discrete values. This result, in turn, means that the design of protection structures can be based on probabilistic methods in which the structure is designed to withstand, for example, an impact of 90%–95% of likely rock fall energies rather than the largest event that may ever occur. Where these large events rarely occur, the cost of providing protection against all rock falls would have to be balanced against the likely consequence of the fall. Methods of probabilistic design and decision analysis for designing protection measures are discussed in Chapter 8.

5.2.1 Theoretical Relationship between Impact Angle and Normal Coefficient of Restitution

The (normal impulse, p_N – relative velocity, v) diagram in Figure 5.4(a) illustrates the influence of impact conditions on the normal coefficient of restitution e_N, keeping in mind the discussion in Section 5.1 that e_N is not a material property, but depends on the impact characteristics. The two dashed lines on Figure 5.4(a) represent changes in the normal velocity during impact for two identical, non-spherical eccentric impacts with respect to mass and impact translation velocity. The approximate relationship between the normal impulse p_N and the normal velocity v_N is assumed to linear with gradient equal to $(1/m)$.

$$v_N = v_{iN} + \frac{p_N}{m} \tag{4.3}$$

---

[*] Because rock fall impacts differ significantly from Newtonian-type impacts (Figure 5.3), the ratio of final to impact normal velocities for rock falls (e_N) may be termed the "apparent" coefficient or restitution. In this book, e_N always refers to rock fall impacts where e_N is not a material property, and can have a value greater than 1.

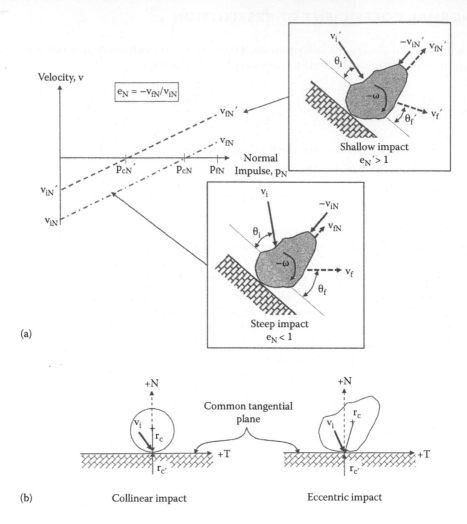

Figure 5.4 Relationship between impact angle and normal coefficient of restitution. a) impulse-relative velocity diagram; b) collinear and eccentric impacts.

In Figure 5.4(a), the only difference between the two lines is the impact angles θ_i, θ_i'. The lines showing the change in normal velocity during impact from a negative value acting toward the slope at the point of impact ($t = i$), to zero at the point of maximum compression ($t = c$), to a positive velocity acting away from the slope at the end of the impact process ($t = f$).

The upper dashed line in Figure 5.4a) represents a shallow angle impact close to the slope surface, while the lower dashed line represents a steeper angle impact ($\theta_i' < \theta_i$). For the shallow impact, the normal velocity component is small relative to the normal component for the steep impact ($v_{iN}' < v_{iN}$). However, the slope of the $[p_N - v]$ line is the same for both impacts because the mass is unchanged; the final normal impulse is also identical.

Examination of the final and initial normal velocities for the two impact conditions in Figure 5.4a) shows that $e_N' > e_N$, and that $e_N' > 1.0$ for the shallow angle impact. A normal coefficient of restitution that is greater than 1 does not mean that energy is created during the impact process; $e_N' > 1$ because the geometry of the impact results in the final normal velocity component being greater than the impact vertical velocity component.

With respect to energy changes during impact, the final velocity will always be less than the impact velocity ($v'_f < v'_i$ and $v_f < v_i$) resulting in a net loss of energy during impact (Giacomini et al., 2010; Buzzi et al., 2012; Asteriou et al., 2012; Spadari et al., 2012). Energy changes during impact are discussed in more detail in Chapter 6.

Figure 5.4(a) also shows that the value of the normal impulse at maximum compression is also influenced by the impact angle ($p_c > p'_c$). That is, a shallow impact results in a "softer" impact than a steeper impact. For steep angle impacts, the body will cause more compression of the slope material, compared to shallow angle impacts, resulting in less rebound and a smaller value of v_{fN}.

The relationship between the impact angle θ_i and e_N shown graphically on Figure 5.4(a) can also be expressed mathematically using the theory of impact mechanics. In developing this relationship it is necessary to distinguish between collinear and eccentric impacts, the geometrical features of which are shown in Figure 5.4(b) where a common tangent plane is established at the contact point that is coincident with the slope surface. The geometry of impact relates to the orientation of the lines r_c and r'_c joining the center of mass of the impacting body with the contact point, as discussed below.

Collinear impact—collinear impact occurs where the line between the center of mass of the impacting body lies on the normal to the common tangent plane; this will occur for a spherical body. For these conditions, the equations of motion for normal and tengential directions can be decoupled such that spin and friction do not contribute to the normal component of the final velocity. That is, the normal coefficient of restitution is equal to the ratio of the final to impact normal velocity components and will be less than 1.0.

Appendix III shows, for collinear impacts, the derivation of equations for the final translational and angular velocities and the restitution angle relative to the slope for conditions where slip stops during impact. These equations can be rearranged to find a mathematical relationship between the final and impact angles, θ_f and θ_i as follows.

The restitution (final, $t = f$) angle θ_f is related to the relative magnitude of the final normal and tangential velocities as shown in Equation (4.24):

$$\tan\theta_f = \left(\frac{V_{fN}}{V_{fT}}\right) \tag{4.24}$$

For a rough, rotating sphere, the final tangential (V_{fT}) and normal velocities (V_{fN}) are given by Equations (4.20) and (4.21) that can be combined with Equation (4.24) as follows:

$$\tan\theta_f = -\frac{V_{iN} \cdot e_N}{V_{iT} - \left(\frac{(V_{iT} + r \cdot \omega_i)}{V_{iN}(1 + r^2/k^2)}\right)} \tag{5.1}$$

Since the impact angle θ_i is related to the impact velocity components as follows

$$\tan\theta_i = \left(\frac{V_{iN}}{V_{iT}}\right)$$

Equations (5.1) and (4.24) can be rearranged as follows to show the relationship between θ_i and e_N :

$$e_N = -\frac{\tan\theta_f}{\tan\theta_i}\left[1 - \frac{1}{(1 + r^2/k^2)}\left(1 + \frac{r \cdot \omega_i}{V_i \cdot \cos\theta_i}\right)\right] \tag{5.2}$$

For an impacting sphere where $k^2 = r^2 \cdot (2/5)$ and $(1+r^2/k^2)^{-1} = 0.29$, Equation (5.2) simplifies to

$$e_N = -\frac{\tan\theta_f}{\tan\theta_i}\left[0.71 - \frac{0.29 \cdot r \cdot \omega_i}{V_i \cdot \cos\theta_i}\right] \tag{5.3}$$

Equations (5.1) to (5.3) are applicable to collinear impact conditions.

Eccentric impact—for actual rock falls where the body is non-spherical and rough, and impact occurs at an oblique angle, eccentric impact conditions apply as defined in Figure 5.4b). For these conditions, the line r_c from the centre of mass of the impacting body does not lie on the on the normal to the common tangent plane, and the equations of motion each involve both normal and tangential forces, such that the effects of friction and normal forces are not separable.

For both collinear and eccentric impact, Goldsmith (1960) provides a graphical solution for finding the final impulses, velocities, and angles. These plots show how the normal and tangential impulses vary throughout the duration of impact and are related to the impact angle and the friction at the contact point. Also, the normal impulse decreases as the impact angle becomes shallower, which is consistent with the impact model shown in Figure 5.4a).

5.2.2 Field Data Showing Relationship between Impact Angle and Normal Coefficient of Restitution

In order to examine the influence of the impact angle θ_i on the normal coefficient of restitution e_N, actual values of θ_i and e_N have been obtained from the field data described in Chapter 2 where impacts and trajectories were carefully documented (Figure 5.5). For each material type and test site, as well as a drop test on concrete (Figure 5.6), the average $[\theta_i - e_N]$ value has been plotted on Figure 5.5. The best-fit line for the average field data points is defined by the following relationship:

$$e_N = 19 \cdot 5\theta_i^{-1.03} \tag{5.4}$$

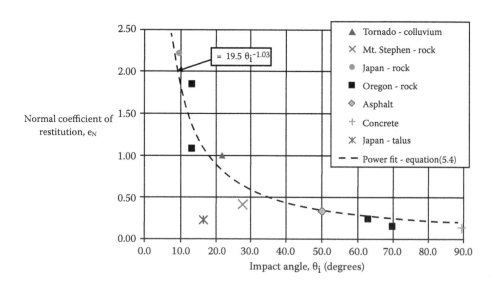

Figure 5.5 Relationship between impact angle θ_i and normal coefficient of restitution e_N with best fit (power) curve for average values of θ_i and e_N each material type.

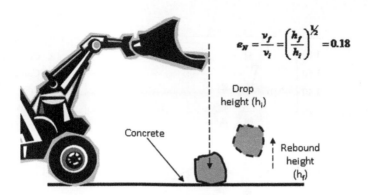

$$e_N = \frac{v_f}{v_i} = \left(\frac{h_f}{h_i}\right)^{\frac{1}{2}} = 0.18$$

Drop
height (h_i)

Concrete

Rebound
height
(h_f)

Figure 5.6 Measurement of normal coefficient of restitution for concrete using drop test to measure rebound height (h_f). (From Masuya, H. et al., 2001. Experimental study on some parameters for simulation of rock fall on slope (h_f). *Proc. 4th Asia-Pacific Conf. on Shock and Impact Loads on Structures*, 63–69.)

Figure 5.5 shows that for steep impacts ($\theta_i > \sim 60$ degrees), e_N is less than about 0.3 and little rebound occurs. Also, for shallow skidding impact of rotating blocks ($\theta_i < \sim 20$ degrees), values of e_N are greater than 1 such that the final normal velocity is greater than the impact normal velocity ($v_{fN} > v_{iN}$). The relationship between θ_i and e_N shown on Figure 5.5 supports the $[p_N - v]$ plot in Figure 5.4a) illustrating the influence of the impact angle on the normal coefficient of restitution.

On Figure 5.5, the reference data point for the field values of e_N is for $\theta_i = 90$ degrees, i.e., a non-rotating block dropped on to a horizontal surface. This is the basic definition of the normal coefficient of restitution as shown in Newton's measurements {Figure 5.3). Figure 5.6 shows a test to measure e_N for concrete in which a block of rock was dropped on a horizontal concrete surface; measurements of the average rebound height showed that $e_N = 0.18$ (Masuya et al., 2001). While similar tests on other slope materials were not carried out, it is reasonable to expect that e_N values for sound rock may be close to 0.18, while impacts on soil and talus would be less than 0.18. In fact, simple observations of dropping rocks on to various ground surfaces shows that rebound heights are always less than about 5 percent of fall height ($e_N \not> 0.2$). This observation is in contrast with e_N values quoted in the literature of about 0.46 for bare rock (*RocScience*, 2003).

It is noted that it would be unusual that a non-rotating rock fall would impact a slope at right angles, and that almost all conditions would tend to have impact angles significantly less than 90°. Therefore, the Newtonian definition for e_N is generally inapplicable to rock falls.

Figures 2.2 to 2.10 show details of the velocities and angles for selected impact points at the various field sites that have been studied to research rock fall characteristics and calculate values for e_N. These sites were selected in order to incorporate a wide variety of both geometric and geologic conditions. That is, the impact angles (θ_i) varied from about 50° at the asphalt site and 70° for the impact in the ditch at the Oregon test site to about 13° for the ¼H:1V face at the Oregon site and 22° on Tornado Mountain. The slope materials studied included rock (Mt. Stephen, Oregon, Ehime), talus (Ehime), colluvium (Tornado Mountain), asphalt, and concrete. The total number of impacts included in this summary is 58.

For each impact at the case study sites, the $[\theta_i - e_N]$ coordinates are plotted on Figure 5.7, with the symbols indicating the slope material. The point on the extreme right side of the plot ($\theta_i = 90°$) is the drop test on concrete. The colluvium points are from the Tornado Mountain site where most θ_i values are in the range of 15° to 25° (see Figure 2.8). The scatter

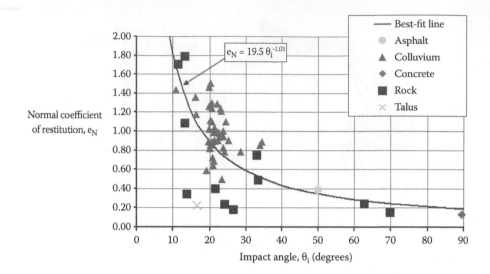

Figure 5.7 Relationship between impact angle θ_i and the normal coefficient of restitution e_N for the rock fall sites described in Chapter 2; total of 58 points for five slope materials.

in the Tornado Mountain points can be attributed primarily to the irregularity in the shape of the falling rock and the variation in the impact angle, because the colluvium was at a uniform slope with no significant roughness. For impacts on talus, two points are available from the Ehime site: impacts #6 and #7 (see Figure 2.6). These two impacts had significantly different behavior with the rock barely leaving the slope after #6, while the trajectory after impact #7 is the longest and highest of the rock fall. As shown in the inserts on Figure 2.6, the values of e_N range from 0.22 for impact #6 to 5.48 for impact #7; impact #6 is plotted on Figure 5.5. For impacts on rock, the range of θ_i values is 11° to 70° and while values of e_N are scattered, they do exhibit a trend with high e_N values at shallow impact angles, to low e_N values at steeper impact angles.

The best-fit curve fitted to the field data on Figure 5.7 shows that θ_i and e_N are related by a power curve defined by Equation (5.4) for which the correlation coefficient R^2 is −0.48. This reasonably high correlation is driven by the points with values of θ_i less than 15 degrees and more than 35 degrees where the scatter is limited compared to points with θ_i in the range of 35 to 70 degrees.

The data plotted on Figure 5.7 shows how the value of e_N decreases as the impact angle θ_i increases from very shallow impacts (low values of θ_i) to a value of 0.18 for normal impacts (θ_i = 90 degrees). For shallow impacts, i.e., θ_i less than about 15 degrees, the value of e_N may be greater than 1. As discussed in Section 5.2.1 above and demonstrated in Figure 5.4, values of e_N greater than 1 does not mean that energy is being created during impact, but only that the impact geometry causes the final normal velocity to be greater than the impact normal velocity. Because the tangential velocity (and energy) always decreases during impact, a net energy loss occurs during impact (see Section 6.1.2).

The symbols on Figure 5.7 show $[e_N - \theta_i]$ values for the five different slope materials listed in the legend. It is noted that values for each slope material are not grouped together, but tend to follow the best-fit line according to the value of the impact angle. This would indicate that e_N is correlated more closely with the impact geometry than with the slope material properties. The values for colluvium occur over a narrow range of θ_i values because they were obtained from the Tornado Mountain site where the slope geometry is uniform, whereas the θ_i values for impacts on rock have a wider range because they were obtained from three sites, each with different geometries.

In Figure 5.7, considerable scatter occurs in the plotted values. This scatter is due to the interaction during the impact process between the irregularly shaped, rotating rock fall and the rough slope surface. The degree of scatter caused by these interactions can be observed on Figure 5.7 for the colluvium impacts. These 43 points are for one ellipsoid-shaped block with a major axis of 1.6 m (5.2 ft) and a minor axis of 1.3 m (4.3 ft), impacting a planar colluvium slope with a slope angle of between 20° and 33° and no significant roughness. Because of the uniformity of the slope, the scatter in the e_N values is almost entirely the result of the attitude and angular velocity of the block as it impacted the slope.

5.2.3 Application of [$\theta_i - e_N$] Relationship to Rock Fall Modeling

One of the input parameters for rock fall modeling programs is the normal coefficient of restitution e_N, for each slope material along the fall path. The program *RocFall 4.0* (*RocScience*, 2012), for example, lists suggested values for e_N that have been obtained by users of the program from their experience of actual rock falls; average values for e_N listed in *RocFall 4.0* are as follows:

Bare rock – 0.46; asphalt – 0.4; soil – 0.34; talus – 0.32

These values listed for e_N are clearly greater than values that would be obtained by dropping blocks of rock on to these surfaces and measuring the rebound height. Figure 5.6, for example, shows a block of rock dropped on to concrete where the measured value of e_N was 0.18. Since the quoted values for e_N have been obtained by back analysis of rock fall events, it is expected that the back analysis values may be consistent with the [$\theta_i - e_N$] relationship shown in Figures 5.5 and 5.7. That is, e_N values of 0.4 to 0.5 correspond to impact angles of between 30° and 40° that are consistent with the values measured in the case studies described in Chapter 2.

The relationship between the impact angle θ_i and the normal coefficient of restitution e_N developed from both field studies and impact mechanics theory shown in Figures 5.4, 5.5, and 5.7 demonstrates that e_N depends more on the impact geometry than on the properties of the slope material. This conclusion is supported by the values for e_N from *RocFall 4.0* listed above that are similar for the four materials, and all are greater than the measured value for concrete.

It is possible that the [$\theta_i - e_N$] relationship given by Equation (5.4) can be used in rock fall modeling where the geometry of the trajectory just prior to impact, in relation to the slope, defines the impact angle θ_i. This value for θ_i can then be input in Equation (5.4) to determine the value of e_N used to calculate, for that impact, final velocities and angles using Equations (4.20) to (4.24). An example of this procedure is shown in the Worked example 4A (Section 4.6.3) where the value of θ_i was 25° and the corresponding value of e_N from Equation (5.4) was 0.71. This value of e_N was then used in Equation (4.21) to calculate the final normal velocity V_{fN}.

5.3 TANGENTIAL COEFFICIENT OF RESTITUTION AND FRICTION

The reduction in tangential velocity during the impact process, as shown on the [$p_N - v$] plot in Figure 4.4, can be quantified in terms of the tangential coefficient of restitution, e_T. That is, e_T is defined as follows:

$$e_T = \frac{v_{fT}}{v_{iT}} \qquad\qquad (4.7)$$

when v_{fT} is the final tangential velocity, and v_{iT} is the impact tangential velocity.

5.3.1 Field Values of Tangential Coefficient of Restitution

The values of e_T measured at the five rock fall locations described in Chapter 2 have been plotted on Figure 5.8 showing a total of 56 impact points for rock (12 points), talus (one point), colluvium (42 points), and asphalt (one point). The legend also shows the average e_T values for impacts on rock and colluvium.

Figure 5.8 shows considerable scatter in the e_T values, as would be expected for the site conditions where the slope surfaces are rough and the rock falls are irregular blocks. In fact, no significant difference is evident in the values of e_T for rock, talus, and colluvium. The lowest value for e_T of 0.24 is that for asphalt, a relatively smooth surface compared to the rock and talus slopes.

The impact process for a rock fall involves the development of shear and normal forces between the rock fall and the slope, and the shear displacement of the rock along this surface. This is typical shearing behavior in which frictional resistance is developed between the two surfaces according to Coulomb's law of friction. Therefore, the tangential coefficient of restitution is analogous to the coefficient of friction μ, and the plot in Figure 5.8 provides an indication of the coefficient of friction values that are developed during impact of rock falls.

The values of e_T plotted on Figure 5.8 appear to be independent of the impact velocity and angle, which is consistent with Coulomb's law in which the friction coefficient is independent of velocity and normal force. The usual practice in rock mechanics is to combine the effects of the frictional properties of the rock material with the roughness, or asperities, of the surface to determine the effective friction angle of a rock surface. If the friction angle of the rock is φ_r and the asperities are simulated as saw-tooth-shaped ridges inclined at angle i, then the effective friction angle of the surface is (Wyllie and Mah, 2002; Patton, 1966):

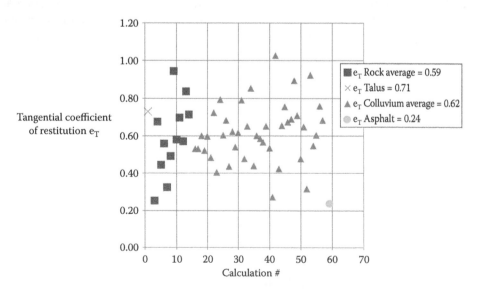

Figure 5.8 Values for tangential coefficient of restitution e_T for 56 impact points at rock fall sites described in Chapter 2.

$$\varphi = (\varphi_r + i) \qquad\qquad (5.5)$$

and

$$\mu = \tan\varphi$$

Equation (5.5) demonstrates that slopes in strong rock with rough surfaces and high values of i can have friction coefficients, μ, that are greater than 1. In rock mechanics, it is usual to assume that the asperities are sheared off as the normal stress increases relative to the rock strength so that the value of i decreases with increasing normal stress. It is possible that similar behavior may occur with rock falls in which fragments of the rock fall break off on impact to form a smoother, more uniform shape with the progression of the fall; the loss of mass during falls is discussed in more detail in Section 6.5. Although the asperities may break down at the impact points, this will not change the basic friction coefficient of the rock that is independent of the normal stress and shearing velocity.

In Figure 5.8, the values of e_T for each field site are plotted as a separate set of points with the "calculation #" referring to successive impact points. The plot shows that the e_T values do not decline during the course of the fall, indicating that loss of mass of the body and the formation of a smoother shape does not result in a significant reduction in the value of e_T.

Furthermore, analysis of correlations for e_T shows no relationship between e_T and the impact angle θ_i. However, a negative correlation exists between e_T and e_N—at high values of e_N values of e_T are low, and as e_N decreases, e_T increases. That is, at shallow impact angles, when the normal restitution velocities can be high, a significant loss of tangential velocity occurs due to friction on the impact surface.

As a comparison with the field values of e_T, direct measurements have been made of the friction coefficient for blocks of rock sliding on various geological materials (Masuya et al., 2001). These tests involved pulling a natural block of rock with a mass of 433 kg (950 lb) on surfaces comprising concrete, a gravel road, and soil (Figure 5.9). The friction coefficient was calculated from the shear force that was recorded by a load cell on the pulling cable, and the shear displacement that was measured with a laser displacement meter. It was found that the average friction coefficients μ for these materials were

$$\mu_{concrete} = 0.59; \ \mu_{gravel} = 0.68; \ \mu_{soil} = 0.90.$$

These measured values of the friction coefficient are comparable to the field values of e_T plotted on Figure 5.8, considering the wide scatter in the results. Also, the lowest values for the friction coefficients are for the smoothest surfaces, asphalt and concrete.

Another application of the coefficient of friction to rock fall behavior is in the calculation of fall velocities. Section 3.2.2 describes how fall velocities are related to the fall height, slope angle, and the effective coefficient of friction of the slope surface. Equation (3.13) relates the fall velocity to these three parameters, and Table 3.1 lists values for effective friction coefficients for a number of slope materials determined from Equation (3.13) for field measurements of actual rock fall velocities.

5.3.2 Application of e_T to Rock Fall Modeling

Reference to Equations (4.20) to (4.24) that define the final velocities and angles for rock fall impacts, shows that the friction coefficient at the impact point does not directly influence the calculated values of these parameters. The final velocity and angle depends on the

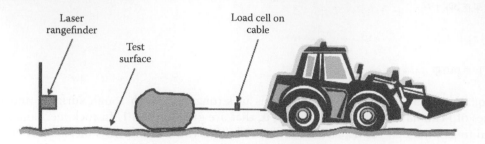

Figure 5.9 Test procedure to measure friction coefficient between block of rock and slope material. (From Masuya, H., et al., 2001. Experimental study on some parameters for simulation of rock fall on slope. *Proc. 4th Asia-Pacific Conf. on Shock and Impact Loads on Structures*, 63–69.)

normal coefficient of restitution and the size (radius, r) and shape of the body (radius of gyration, k).

As discussed in Chapter 6 on energy losses during impact, the friction coefficient is a component of the equations defining the loss of energy for a rotating body where energy is lost due to friction during the slip phase of the impact process.

Chapter 6

Energy Changes during Impacts and Trajectories

A rock fall event comprises a series of impacts, each followed by a trajectory. When the velocity and energy that are lost during impact are less than the velocity and energy that are gained during the subsequent trajectory, the rock fall will continue down the slope. However, as the slope angle decreases such that the impacts occur at a steeper angle with the slope surface and the trajectories become shorter, then the energy lost at the point of impact will be greater than the energy gained during the trajectory and the rock fall will come to a stop. For example, the relationship between slope geometry and energy loss is illustrated in Figure 6.1 where a high-velocity rock fall on a steep rock face was stopped in a short distance when it impacted a level bench; no significant damage occurred to the building.

A similar situation is shown in Figure 2.8 for the Tornado Mountain case study where the rock fall continued for a slope distance of about 700 m (2,300 ft) where the slope angle was uniform at 30° to 22°, with little reduction in velocity. However, once the rock impacted the level ditch beside the railway, about 70% of the energy was lost, and it stopped within a distance of about 30 m or 100 ft (see Figure 9.1).

For any rock fall, each impact and trajectory will be different as the result of variations in the slope properties, slope angle, material type and roughness, and the attitude of the rock at the impact point. These differences in site conditions result in corresponding differences in the translational and rotational velocities at each impact point. Regardless of these velocity variations, the energy changes that occur during a fall comprise a reduction in kinetic energy during impact as a result of compression and friction, followed by an increase in kinetic energy during the trajectory as a result of gravitational acceleration. The rotational energy will also change during impact but will remain constant during the trajectory.

This chapter discusses changes in the translational and rotational energies that occur during rock falls, and how they can be quantified. This information can be used in the design of rock fall containment structures, such as barriers and fences, with respect to both their location and allowable impact energy capacity. A technique is demonstrated (Section 6.4) in which the potential, kinetic, and rotational energies can be partitioned and then plotted for every stage of the rock fall. Such a plot will indicate the lowest energy location along the rock fall path, and the optimum location for the barrier or fence.

6.1 IMPACT MECHANICS THEORY AND KINETIC ENERGY CHANGES

This section shows the development of equations defining the changes in kinetic energy that occur during impact. Two cases are considered: first, a nonrotating body impacting the slope in the normal direction and, second, a rotating body impacting the slope at an oblique angle such that energy changes occur in both the normal and tangential directions.

Figure 6.1 Rock fall that stopped, just before causing serious damage to a building, when it impacted a horizontal surface that absorbed most of the fall energy.

6.1.1 Kinetic Energy Changes for Normal Impact, Nonrotating Body

The theory of impact mechanics addresses the normal force generated in an infinitesimal, deformable particle at the contact point (Figure 6.2(a)). As discussed in Chapter 4 (Section 4.1), it was demonstrated that the impact comprises two phases—a compression phase up to the point of maximum compression (impulse = p_{cN}), followed by a restitution phase from the point of maximum compression to the point of separation (impulse = p_{fN}; Figure 6.2(b)). In terms of the kinetic energy changes during impact, energy is absorbed by both the rock and the slope material during compression, and then a portion of this energy is recovered during restitution. The recovered elastic strain energy is converted into kinetic energy, and it is this energy that drives the rock away from the slope. For perfectly elastic materials, no energy is lost during impact, while for perfectly plastic materials all the impact energy is absorbed in compression, and the rock fall remains in contact with the slope because no energy is recovered to produce separation (see Figure 4.3).

The process of energy loss and recovery during impact can also be expressed in terms of [deformation, δ – normal force, F] plots, and [normal impulse, p_N – relative velocity, v] plots as shown in Figure 6.2. In Figure 6.2(b), the deformation at maximum compression is δ_c, while the deformation at the completion of the impact is δ_f, where $\delta_c > \delta_f$ because, for a partially elastic impact, only part of the deformation is recovered during the restitution phase. The energy associated with each phase of the impact is equal to the area under curve on the [δ – F] plot, with the energy lost during normal compression being E_{cN}, and the energy recovered during normal restitution being $(E_{fN} – E_{cN})$ (Figure 6.2(b)). On the [p_N – v] plot, these energies are equal to the triangular areas for the impulse at maximum compression p_{cN} and the final impulse p_{fN} (Figure 6.2(c)).

Equations for energy changes during impact can be developed from the [p_N – v] plot illustrated in Figure 6.2(c), with the impact process simulated by the infinitesimal deformable

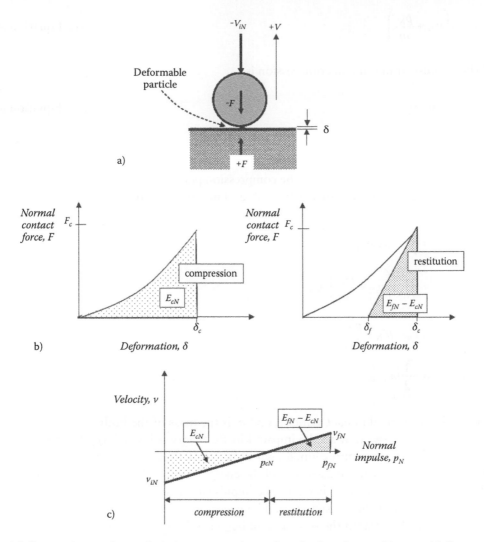

Figure 6.2 Energy changes (normal) during compression and restitution phases of impact. (a) Forces generated at contact point during normal impact; (b) energy plotted on [force, F-deformation, δ] graph; (c) energy changes plotted on [normal impulse, p_N-relative velocity, v] graph.

particle at the impact point. During impact, the energy E_N generated in the particle by the normal component of the force F_N can be calculated from the relationship between the force and the differential normal impulse: $dp_N = F_N\ dt = m\ dv$ (see Equation 4.1a)), so that the energy generated from the moment of impact ($p_N = 0$) up to time t and impulse p_N is

$$E_N = \int_0^t F_N \cdot v\ dt = \int_0^{p_N} v\ dp_N \tag{6.1}$$

For the compression phase of the impact up the impulse p_{cN}, the relationships between impulse and velocity are as follows:

$$v_N = \left(v_{iN} + \frac{p_N}{m} \right)$$

(see Equation (4.3))

and the impulse at maximum compression is

$$p_{cN} = -m \cdot v_{iN}$$

(see Equation (4.4))

where v_{iN} is negative because it acts toward the slope in the direction of the $(-N)$ axis (Figure 6.2(a)).

Therefore, the energy lost during the compression phase of the impact is given by the area on the $[p_N - v]$ plot between impact ($p_N = 0$) and maximum compression ($p_N = p_{cN}$):

$$\begin{aligned}
E_N(p_{cN}) &= \int_0^{p_{cN}} v_N\, dp_N \\
&= \int_0^{p_{cN}} \left(v_{iN} + \frac{p_N}{m} \right) dp_N \\
&= v_{iN} \cdot p_{cN} + \frac{p_{cN}^2}{2m} \\
&= -\frac{1}{2} m \cdot v_{iN}^2
\end{aligned}$$

(6.2)

where v_{iN} is the normal impact velocity, and m is the mass of the body.

Equation (6.2) shows that all the impact kinetic energy is lost ($E_N(p_{cN})$ is negative) up to the point of maximum compression, δ_c, when the normal velocity is reduced to zero ($v_N = 0$).

A similar approach can be used to find the energy recovered during the restitution phase of the impact ($E_N(p_f) - E_N(p_c)$) between the impulse at maximum compression (p_{cN}) and the impulse at the end of the impact (p_{fN}). The energy recovered, which is termed the elastic strain energy, is the area on the $[p_N - v]$ plot between these two impulses:

$$\begin{aligned}
E_N(p_{fN}) - E_N(p_{cN}) &= \int_{p_{cN}}^{p_{fN}} \left(v_{iN} + \frac{p_N}{m} \right) dp_N \\
&= \left[v_{iN} \cdot p_{fN} + \frac{p_{fN}^2}{2m} \right] - \left[v_{iN} \cdot p_{cN} + \frac{p_{cN}^2}{2m} \right] \\
&= \frac{m \cdot v_{iN}^2}{2} \left(1 - \frac{p_{fN}}{p_{cN}} \right)^2
\end{aligned}$$

(6.3)

Derivation of this equation is shown in Appendix I, Equation (I.9). Alternatively, the elastic strain energy can be calculated from the area of the restitution triangle between impulse values p_{cN} and p_{fN} on Figure 6.1(c)). Derivation of Equation (6.3) from the area of this triangle is also shown in Appendix III, Equation (III.3).

Equations (6.2) and (6.3) together define the net energy loss during normal impacts as:

$$E_N(net) = [\text{energy lost in compression}] + [\text{energy gained in restitution}] \qquad (6.4)$$

$$= \left[E_N(p_{cN}) \right] + \left[E_N(p_{fN}) - E_N(p_{cN}) \right]$$

$$= -\frac{m \cdot v_{iN}^2}{2} + \frac{m \cdot v_{iN}^2}{2} \left(1 - \frac{p_{fN}}{p_{cN}} \right)^2$$

$$= -\frac{m \cdot v_{iN}^2}{2} \left[1 - \left(1 - \frac{p_{fN}}{p_{cN}} \right)^2 \right]$$

Equation (4.12) defines the relationship between normal impulses p_{fN}, p_{cN} and the normal coefficient of restitution e_N, as:

$$p_{fN} = -m \cdot v_{iN}(1 + e_N) = p_{cN}(1 + e_N) \qquad (4.12)$$

and

$$\frac{p_{fN}}{p_{cN}} = (1 + e_N)$$

and

$$e_N^2 = \left(1 - \frac{p_{fN}}{p_{cN}} \right)^2$$

Substitution of Equation of (4.12) into Equation (6.4) gives the following expression for the net energy loss during normal impact:

$$E_N(net) = -\frac{1}{2} m \cdot v_{iN}^2 (1 - e_N^2) \qquad (6.5)$$

In the development of Equation (6.5) for normal impact, the value of the normal coefficient of restitution, e_N is always less than 1 because it is defined by the energy losses for a non-rotating body. In contrast, Figures 5.5 and 5.7 show the conditions that result in $e_N > 1$, that is, when a rotating body impacts the slope at a shallow angle, $\theta_i < 15°$ approximately. Therefore, for shallow impact of a rotating, rough body, e_N may be greater than 1, but energy will be lost for the overall impact because energy is also lost in the tangential component of impact.

Worked example 6A—energy loss for normal impact: as an illustration of the application of Equation (6.5), consider the test to determine the value of e_N for concrete shown in Figure 5.6. If the block of rock is dropped vertically on to the concrete from a height of 5 m, then $v_{iN} = \sqrt{(2 \, g \, h)} = 9.9 \, \text{m} \cdot \text{s}^{-1}$. The tests described in Figure 5.6 show that $e_N = 0.18$ for these conditions, so that $(1 - e_N^2) = 0.97$. If the block mass is 1000 kg, then the impact kinetic energy is $\frac{1}{2} \, m \cdot v_{iN}^2 = 49 \, \text{kJ}$, and the net energy loss, $E_N(net) = -49 \times 0.97 = -47.5 \, \text{kJ}$, or 97% of the impact energy.

In energy calculations, it is necessary to use the mass and not the weight of the body. That is, the unit weight of rock is generally equal to 26 kN · m⁻³, or I cu. m. of rock weighs 26,000 N. Since weight is a force, the mass is related to the weight by the gravitational acceleration, g = 9.81 m · s⁻², and the mass of I cu. m. of rock is 26,000/9.81 = 2,650 kg.

6.1.2 Kinetic Energy Changes for Inclined Impact, Rotating Body

The usual impact condition for a rock fall is an oblique impact of the body with the slope, with the body rotating in the negative direction, that is, the peripheral velocity $(-r \cdot \omega_i)$ is in the direction opposite to the positive tangential axis. Calculation of the energies lost during compression, and gained during restitution, uses the same principles as shown in Figure 6.2, and described in Section 6.1.1 for normal impact of a nonrotating body. However, for oblique impact of a rotating body, energy losses occur to the normal and tangential components of kinetic energy, as well as the rotational energy.

The kinetic energy changes are separated into the normal and tangential components, with the normal component comprising loss of energy up to the point of maximum compression, δ_c followed by partial recovery during restitution at the final compression δ_f $(\delta_f < \delta_c)$. In the tangential direction, energy is lost throughout the impact process as the result of friction acting at the contact point defined by the friction coefficient μ. These energy changes can then be combined to find the net kinetic energy loss during impact.

The energy calculations assume that the direction of slip is constant throughout the impact, which occurs if the initial slip velocity speed (v_{is}) is large so that slip does not halt during contact, or the initial slip speed is zero. For rock falls, this assumption is valid because of the high friction at the impact points combined with the generally high tangential velocity component that generates spin in the negative direction as demonstrated in Figure 6.3.

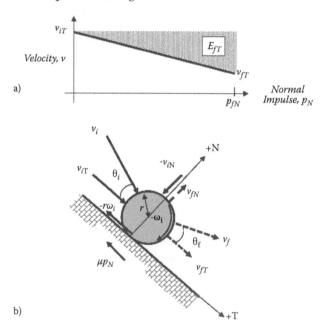

Figure 6.3 Reduction in tangential velocity, v_T during impact. (a) $[p_N - v]$ diagram showing change in v_T during impact, and corresponding reduction in energy, E_T; (b) changes in velocity components.

Normal component of energy loss—The equation defining the normal velocity v_N at any time during oblique impact of a rotating body is given in Appendix II, which shows that v_N is related to inertial coefficients β, the friction coefficient μ, and the mass of the body m:

$$v_N = v_{iN} + \frac{(\beta_3 + \mu \cdot \beta_2)p_N}{m} \tag{II.15b}$$

where inertial coefficients β_2 and β_3 are as defined in Appendix II.1, Equations (II.11); inertial coefficients are defined by the size (radius, r) and shape (radius of agration, k) of the rotating body.

At the point of maximum compression when the body is momentarily stopped, $v_N = 0$ at impulse p_c (Figure 6.2c)). Therefore,

$$p_{cN} = \frac{-m \cdot v_{iN}}{(\beta_3 + \mu \cdot \beta_2)} \tag{6.6}$$

The partial energy lost during the period of compression from the time of impact ($p_N = 0$) up to impulse at maximum compression ($p_N = p_{cN}$) is determined by integration of the area on the $[p_N - v]$ plot:

$$E_N(p_{cN}) = \int_0^{p_{cN}} v_N \, dp_N \tag{6.7}$$

$$= \int_0^{p_{cN}} \left(v_{iN} + \frac{(\beta_3 + \mu \cdot \beta_2)p_N}{m} \right) dp_N$$

$$= \left[v_{iN} \cdot p_{cN} + \frac{(\beta_3 + \mu \cdot \beta_2)p_{cN}^2}{2m} \right] - [0]$$

$$= \frac{-m \cdot v_{iN}^2}{(\beta_3 + \mu\beta_2)} + \left(\frac{(\beta_3 + \mu \cdot \beta_2)}{2m} \right) \left(\frac{-m \cdot v_{iN}}{(\beta_3 + \mu \cdot \beta_2)} \right)^2$$

$$= \frac{-m \cdot v_{iN}^2}{(\beta_3 + \mu \cdot \beta_2)}$$

The same approach can be taken to find the partial energy recovered during restitution, determined by integration of the area on the $[p_N - v]$ plot between impulses p_{cN} and p_{fN}:

$$E_N(p_f) - E_N(p_c) = \int_{p_{cN}}^{p_{fN}} v_N \, dp_N$$

$$= \int_{p_{cN}}^{p_{fN}} \left(v_{iN} + \frac{(\beta_3 + \mu \cdot \beta_2)p_N}{m} \right) dp_N$$

$$= \left[v_{iN} \cdot p_{fN} + \frac{(\beta_3 + \mu \cdot \beta_2)p_{fN}^2}{2m} \right] - \left[v_{iN} \cdot p_{cN} + \frac{(\beta_3 + \mu \cdot \beta_2)p_{cN}^2}{2m} \right]$$

$$= v_{iN}(p_{fN} - p_{cN}) + \left[\frac{(\beta_3 + \mu \cdot \beta_2)}{2m} \right] (p_{fN}^2 - p_{cN}^2)$$

From Equations (II.15b) and (4.4)

$$v_{iN} = -\frac{(\beta_3 + \mu \cdot \beta_2)}{m} p_{cN}$$

Therefore,

$$E_N(p_{fN}) - E_N(p_{cN}) = \frac{(\beta_3 + \mu \cdot \beta_2)}{2m}(p_{fN} - p_{cN})^2 \qquad (6.8)$$

The net normal kinetic energy loss during impact is found by adding Equations (6.7) and (6.8) where

$E_N(net)$ = [energy lost in compression] + [energy gained in restitution]

Tangential component of energy loss—Because shearing occurs during the rock/slope contact for an oblique impact, a frictional force is generated at this point. The tangential component of the reaction force is in the opposite direction to the slip direction, and the tangential velocity reduces from the impact value v_{iT} to a final velocity of v_{fT} as shown in Figure 6.3(a). The uniform velocity reduction during impact assumes that the body slips throughout the impact (see Section 4.5). This reduction in velocity results in a loss of energy during the impact defined by the triangular area on Figure 6.3(a) follows:

$$E_T(p_f) = \int_0^{p_f} v_T \, dp$$

According to Coulomb's definition of friction, the coefficient of friction μ is the ratio of the tangential to normal impulses or forces acting at the contact point, and the relationship between the impulse components is as follows:

$$dp_T = -\mu \, dp_N \qquad (4.14a)$$

The negative sign denotes that the friction force acts in the opposite direction to the positive tangential impulse.

The equation defining the tangential velocity v_T at any time during oblique impact of a rotating body is given in Appendix II, which shows that v_T is related to inertial coefficients β_1 and β_2, the friction coefficient μ, and the mass of the body m as follows:

$$v_T = v_{iT} - \frac{(\beta_2 + \mu \cdot \beta_1)p_N}{m} \qquad (II.15a)$$

where inertial coefficients β_1 and β_2 are as defined in Section II.1, Equations (II.11); inertial coefficients are defined by the size (r) and shape (k) of the rotating body.

The energy lost by the tangential component of impulse is

$$(6.9)$$

$$E_T(p_{fN}) = -\mu \int_0^{p_{fN}} v_T \; dp_N$$

$$= -\mu \int_0^{p_{fN}} \left(v_{iT} - \frac{(\beta_2 + \mu \cdot \beta_1)}{m} p_N \right) dp_N$$

$$= -\mu \left[v_{iT} \cdot p_{fN} - \frac{(\beta_2 + \mu \cdot \beta_1)}{2 \, m} p_{fN}^2 \right]$$

$$= -\frac{\mu \cdot p_{fN}}{2} \left(2v_{iT} - \frac{(\beta_2 + \mu \cdot \beta_1)}{m} p_{fN} \right)$$

The net kinetic energy loss during impact, comprising normal and tangential energy components, can be found by adding Equations (6.7), (6.8), and (6.9).

Worked example 6B—energy loss for oblique impact: as an illustration of the application of energy loss Equations (6.7) to (6.9), consider the case study of the rock falls on Tornado Mountain as described in Section 2.2.1 with the impact and restitution velocity components shown in Figure 2.8 for impact #A26. The boulder was assumed to be an ellipsoid with semi-major axes as follows:

$a = 0.8$ m; $b = 0.65$ m and $c = 0.65$ m, where the body is rotating about axis c.

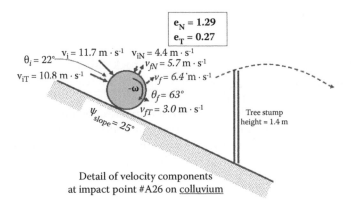

Figure 6.4 Translational and angular velocity components at impact point #A26 for Tornado Mountain rock fall event.

The volume of the ellipsoid is 1.42 cu. m., and the mass is about 3,750 kg if the rock unit weight is 26 kN $\cdot$ m^{-3}. The radius of gyration of the body k is $((a^2 + b^2)/5)^{1/2} = 0.46$ m. The average radius of the body is $(a + b)/2 = 0.725$ m.

The values of the three inertial coefficients are given by Equation (II.11) as follows:

$\beta_1 = 1 + r_b^2/k^2 = 3.00$; $\beta_2 = r_a \cdot r_b/k^2 = 2.46$; $\beta_3 = 1 + r_a^2/k^2 = 4.02$

For an impact velocity of v_i = 11.7 m · s^{-1} at an angle θ_i = 22°, the normal and tangential components of the initial velocity are

$$v_{iN} = -11.7 \sin(22) = -4.4 \text{ m} \cdot \text{s}^{-1} \text{ and } v_{iT} = 11.7 \cos(22) = 10.8 \text{ m} \cdot \text{s}^{-1}$$

For the granular colluvium on the slope surface at the impact point, the friction coefficient μ can be assumed to have a value of 0.68, as determined in the laboratory testing carried out by Masuya et al. (2001), and described in Section 5.3.

Substitution of these parameter values in Equation (6.6) gives a value for the impulse at maximum compression, p_{cN} = 2898 kg · m · s^{-1}. The calculated value of the normal coefficient of restitution e_N at impact #A26 is 1.30 (Figure 2.8), so the final impulse, $p_{fN} = p_{cN} (1 + e_N)$ = 6,665 kg · m · s^{-1}.

Using these values for impulses p_{fN} and p_{cN}, the changes in energy at impact #A26 are calculated as follows:

$$E_N(p_c) = -12.9 \text{ kJ}; \quad E_N(p_f) - E_N(p_c) = 10.7 \text{ kJ}; \quad E_T(p_f) = -31.3 \text{ kJ}$$

and the net energy loss during impact due to changes in translational velocities is equal to the sum of these energy changes, ΔE = is −33.5kJ.

6.2 ROTATIONAL ENERGY GAINS/LOSSES

The rotational energy E_r of a falling body depends on its mass and shape as defined by its moment of inertia I, and the angular velocity ω, as follows:

$$E_r = \frac{1}{2} I \cdot \omega^2 \tag{6.10}$$

As an example of measured angular velocities of rock falls, Section 3.4.1 (Figure 3.10) describing the Ehime test site in Japan shows that ω first increases with the fall distance and then tends to a terminal velocity in the range of about 15 to 35 rad · s^{-1}. Furthermore, the angular velocity may increase or decrease during impact, depending on the geometry of the impact, so the rotational energy may also increase or decrease during impact.

Based on Equation (6.10), where the angular velocity changes from ω_i at the point of impact to ω_f at the termination of impact, the change in rotational energy during impact is

$$\Delta E_r = -\frac{1}{2} I (\omega_i^2 - \omega_f^2) \tag{6.11}$$

6.3 TOTAL ENERGY LOSSES

The total energy loss during impact is the sum of the translational and rotational energy losses. Referring to the calculated energy losses for Tornado Mountain impact #A26 that are calculated in Worked examples 6B and 6C above, the total energy loss is equal to:

$$\Delta E = \Delta KE + \Delta RE = (-33.5) + (-18.3) = -51.8 \text{ kJ}$$

Worked example 6C—change in rotational energy: as an illustration of the application of Equations (6.10) and (6.11), consider rock fall impact #A26 at Tornado Mountain discussed in the previous section. This body was assumed to be an ellipsoid with semi-axes of $a = 0.8$ m, $b = 0.65$ m, and $c = 0.65$ m, and mass, m, of approximately 3,750 kg. The moment of inertia of an ellipsoid rotating about axis c is (see Table 4.1):

$$I = \frac{m}{5}(a^2 + b^2) = 797 \text{ kg} \cdot \text{m}^2 \tag{6.12}$$

No measurements of the angular velocity are available for this fall. However, an approximate value for the angular velocity can be obtained from Equation (3.17) where a relationship between the angular velocity ω, the final translational velocity v_0, and the radius r of the body has been developed from the measurements of the angular velocity of test rocks. That is, at the start of the trajectory where the angular velocity is ω_0:

$$\omega_0 \approx \frac{v_0}{r} \approx \frac{v_f}{r} \text{ (see Figure 3.3)} \tag{3.17}$$

If the radius $r = 0.725$ m, and the final translational velocity, $v_{f(A26)} = 6.4$ m $\cdot$ s^{-1} (Figure 2.8), then the final angular velocity, $\omega_{f(A26)} = -8.8$ rad $\cdot$ s^{-1}.

The impact angular velocity at #A26, $\omega_{i(A26)}$ is equal to the final angular velocity at impact point #A25, equal to $\omega_{i(A25)}$, because the angular velocity does not change during the trajectory from #A25 to #A26. Trajectory calculations for the rock fall show that the final velocity at #A25 was $v_{f(A25)} = 8.07$ m $\cdot$ s^{-1}, and from Equation (3.17), $\omega_{i(A26)} = -11.1$ rad $\cdot$ s^{-1}. Based on these translational velocity parameters, the angular velocity decreased from -11.1 to -8.8 rad $\cdot$ s^{-1} during impact #A26. From Equation (6.11), the change in rotational energy during the impact is

$$\Delta E_r = -\frac{1}{2}797\left((-11.1)^2 - (-8.8)^2\right)$$

$$= -18.3 \text{ kJ}$$

These calculations show that the rotational energy loss is about 35% of the total energy loss; the field data (Chapter 2) shows that this ratio between the kinetic and rotational energies is typical for rock falls.

The theoretical energy losses calculated using impact mechanics as described in Sections 6.1 and 6.2 and demonstrated in Worked examples 6A, 6B, and 6C, can be compared with the change in kinetic energy, ΔKE calculated from the initial and final velocities for impact #A26 as shown in Figure 6.4 where $v_i = 11.7$ m $\cdot$ s^{-1} and $v_f = 6.4$ m $\cdot$ s^{-1}.

$$\Delta KE = -\frac{1}{2}m(v_i^2 - v_f^2) = -\frac{1}{2}3750(11.7^2 - 6.4^2) = -180 \text{ kJ}.$$

These values for the theoretical and field net energy losses show that the theoretical value calculated from impact mechanics theory is less than the actual energy losses. That is, for the translational velocities, the theoretical energy loss is only 19% ($-33.5/-180 = 0.19$) of the actual energy loss, or for the combined kinetic and rotational energies, the theoretical energy loss is about 29% ($-51.8/-180 = 0.29$) of the field value. It is expected that this difference is because the impact mechanics theory does not fully account for the plasticity that occurs in actual rock fall impact, or the loss of mass due to fragmentation of the block (see Section 6.5).

For impacts in the other case studies described in Chapter 2, the theoretical energy losses are also less than the actual energy losses, with the differences being similar to those for Tornado Mountain impact #A26. Furthermore, the differences between the theoretical and actual energy losses shows no apparent relationship to the type of slope material, although further studies of the available data may provide information on a factor to be included in the energy loss equations to account for the plastic deformation at the impact points.

6.4 ENERGY LOSS DIAGRAMS

One of the design requirements of rock fall protection structures is locating the structure where the rock fall energy is at a relatively low value along the rock fall path. The general behavior of rock falls with respect to energy changes during the series of impacts and trajectories is as follows:

- Potential energy (*PE*) will decrease continuously during the fall in proportion to the loss of elevation;
- Translational kinetic energy (*KE*) will decrease during impact in proportion to the reduced velocity, as a result of compression of the slope material in the normal direction, and frictional losses in the tangential direction;
- Rotational energy (*RE*) may decrease or increase during impact, depending on the attitude of the body at the point of impact (see Figure 3.11);
- Translational energy will increase during the trajectory phases of the fall as the result of gravity acting on the body to increase the vertical velocity component;
- Rotational energy will remain constant during trajectories because no rotational forces act on the body when it is moving through the air.

It is assumed that no energy losses due to air friction occur during trajectories because they would be very small compared to impact energy losses. Also, gravity effects during impacts can be ignored because reaction forces are large compared to the body force and act for a very short time, and no significant change in the position of the body occurs during impact.

These energy changes during a rock fall can be illustrated on energy loss diagrams, of which two types are discussed in the following sections.

6.4.1 Energy Partition Diagram for Potential, Kinetic, and Rotational Energies

At the moment that a rock fall detaches from the slope, it will possess potential energy equal to

$$PE = m \cdot g \cdot h \tag{6.13}$$

where *m* is the mass of the body, *g* is the gravitational acceleration, and *h* is the total fall height. As the body falls during the first trajectory, the potential energy is converted into kinetic energy that increases as the body accelerates. At the first impact point, the velocity decreases as the slope deforms and the body slips, causing the kinetic energy to decrease. Also at the impact point, the body starts to rotate and gain rotational energy as the result of friction forces acting on the periphery of the body. Once the body leaves the slope, the translational velocity and kinetic energy increase due to gravitational acceleration, while the angular velocity remains constant. This sequence of events is repeated with each impact.

Figure 6.5 illustrates an energy partition diagram for a cubic concrete block at the Ehime test site in which the *PE, KE, RE,* are all represented at each every impact point and during each trajectory. The vertical axis shows the overall vertical height of the fall from the source, with the elevation of each impact point defined. The horizontal axis shows the partition of the energy throughout the fall, with the maximum value equal to the potential energy at the source; the potential energy decreases continually during the fall. At each impact point, the decrease in velocity, and corresponding kinetic energy loss, is represented by the decrease in the width of the *KE* area. The rotational energy area remains constant during trajectories, but for this test, the rotational energy increases with each impact as shown by the increase in width of the area representing *RE* during impact. The relative widths of the *KE* and *RE* bands indicate the low ratio between *RE* and *KE*.

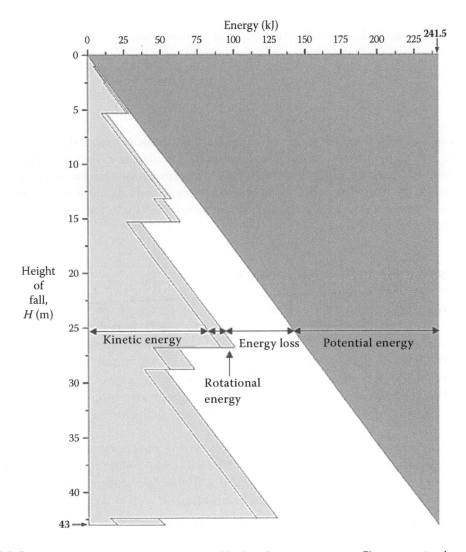

Figure 6.5 Energy loss diagram for cubic, concrete block with constant mass at Ehime test site, Japan. See Figure 2.6. (Ushiro et al., 2001.)

The area between the potential energy area and the combined kinetic-rotational area represents the cumulative energy loss at the impact points that becomes wider with each impact. All the energy has been dissipated when the rock stops moving.

The value of the energy partition diagram is that it enables low energy locations on the slope to be identified that are the optimum locations for rock fall fences or barriers. That is, energy will be high on the steep, upper part of the face but will diminish as the slope angle is reduced. Also, the trajectories will be close to the ground on the lower slope, which will limit the height of protection structures.

Another factor to consider in the location of fences or barriers is to find areas immediately after an impact point where the kinetic energy has been lost during impact and before gravity has caused the velocity to increase during the next trajectory. Impact points will often occur on benches where rock falls accumulate.

6.4.2 Energy Head

An alternative method of showing the progressive energy loss for a series of impacts is to plot the "energy head," given by

$$E = z + \frac{v^2}{2g} \tag{6.14}$$

where z is the body elevation and v is the length of the velocity vector (Hungr and Evans, 1988). A plot of energy head comprises a horizontal line during each trajectory where the energy head is constant and a drop in the energy head representing the loss of energy at each impact point. Plots of this type have been used to show the transition from trajectory to rolling phase of rock falls where the slope of the energy line equals the coefficient of rolling friction for the slope material.

6.5 LOSS OF MASS DURING IMPACT

The energy change calculations discussed previously in this chapter all assume that the mass of the rock fall remains constant throughout the fall. In reality, falls will break up to some extent at the impact points so that the final mass will be less than the mass at the source. Possible reasons for breakup of the body are that the initial block contains planes of weakness along which fracture can readily occur during impact with the slope. Also, the initial body may have an irregular form that will progressively break down into a more uniform spherical or ellipsoidal shape that will spin and bounce more readily than an elongated shape.

Quantitative information on the loss of mass during the course of a rock fall can be obtained from studies carried out in Italy (Nicolla et al., 2009), and from the Tornado Mountain rock fall site described in Section 2.2.1. The Italian studies involved two natural rock fall sites where previous events could be observed, as well as a test site where the falls were carefully documented. These data have been used to develop a relationship between the body size and the distance traveled from the source.

A summary of the two Italian rock fall sites and the Tornado Mountain case study described in Chapter 2 is a follows:

- **Camaldoli Hill, Naples**—The slope geometry at the rock fall locations comprised a 200 m (660 ft) high upper rock slope at an overall angle of about 56°, above a 60 m (200 ft) high talus slope at an angle of 37°. The maximum horizontal fall distance was about 200 m (660 ft). The rock forming the cliff is a tuff (Neapolitan yellow tuff) of varying composition that had a uniaxial compressive strength up to about 10 MPa and deformation modulus of about 10 GPa. The rock contains sets of subvertical joints that are parallel and normal to the slope, and a subhorizontal joint set that dips at 5° to 38° and forms the base of columnar blocks. The joint spacing has a maximum value of about 3 to 5 m (10 to 15 ft), which defines the maximum block size.

  Block sizes were assessed by mapping the dimensions of blocks in-place on the cliff face, and of fallen blocks on the slope. A total of 298 blocks were mapped on the face with volumes ranging from 1 cu. m (1.3 cu. yd) to 50 cu. m (65 cu. yd), with about 34% of the blocks having volumes less than 2.5 cu. m (3.3 cu. yd). A total of 120 fallen blocks were mapped, of which 96% had volumes less than 5 cu. m (6.6 cu. yd) indicating the fragmentation of blocks during falls.

  In addition to the natural falls, a test was conducted on a slope with similar geometry where images of the falls were analyzed to determine the approximate size of blocks before and after impact. The initial volumes of the test blocks ranged from 1 to 12 cu. m (1.3 to 16 cu. yd).

- **Monte Pellegrino, Palermo**—The slope geometry comprised a 100 m (330 ft) high slope at a face angle of about 42°, gradually flattening to a slope angle of 23°; the maximum vertical fall height was 290 m (950 ft) and the horizontal distance was about 300 m (980 ft). The rock forming the cliffs is a strong, blocky limestone with a compressive strength of about 100 MPa and a modulus of about 100 GPa.

  Similarly to the Camaldoli Hill site, block sizes were measured *in situ* and on the run-out area. Most of the *in situ* blocks had volumes up to 2.5 cu. m (3.3 cu. yd), with the largest volume being about 27 cu. m (35 cu. yd), while 87% of the fallen blocks had volumes of less than 5 cu. m (6.6 cu. yd).

- **Tornado Mountain, Canada**—At this site where the two separate rock fall paths were clearly distinguishable on the slope, it was possible to measure the final dimensions of the rock fall, as well as fragments that had broken from the block at specific impact locations (see Figure 2.7). It was not possible to access the source area on a steep rock face to measure the dimensions of the original rock falls. However, from observations of the size of the clearly visible source area, and of the very strong, massive limestone forming the slope it is estimated that the original rock falls had volumes in the range of 30 to 50 cu. m (40 to 65 cu. yd).

The Italian field data were used to develop the following relationship between the volume of the source rock fall Ω_0, and the volume Ω at any horizontal distance x from the source:

$$\frac{\Omega}{\Omega_0} = \frac{1}{(1+x\lambda)} \tag{6.15}$$

where λ (m^{-1}) is a reduction coefficient defining the loss of mass with distance fallen.

Figure 6.6 shows the values of the coefficient λ defining the envelopes of loss of mass data for the Mt. Pellegrino and Camaldoli Hill rock fall sites. For the Mt. Pellegrino data, the values of λ are in the range of 0.0035 and 0.01 m^{-1}, while for the Camaldoli Hills, the values of λ values are in the range of 0.008 to 0.02 m^{-1}. These results demonstrate that as much as 80% of the original rock mass is lost during the course of the rock fall, with most of this

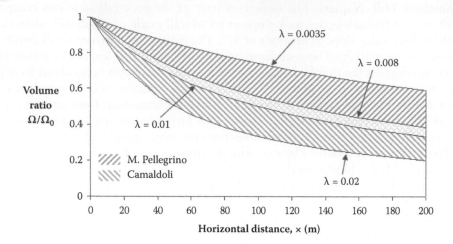

Figure 6.6 Plot of horizontal rock fall distance *x* against loss of volume ratio, Ω/Ω_0 showing ranges of values of λ for data from Mt. Pellegrino and Camaldoli Hills. (From Nicolla, N. et al., 2009. *Rock Mech. Rock Eng.*, 42, 815–833.)

loss probably occurring in the first few impacts where the rock is falling fast down the steep face, and the irregular shape is broken down at the impact points. Also, the initial impacts may be on the rock face where more fragmentation may occur compared to the talus or soil lower on the slope.

It is of interest that the limestone at Mt. Pellegrino is an order of magnitude stronger than the tuff in the Camaldoli Hills and that the reduction in mass is less at Mt. Pellegrino than at the Camaldoli Hills. These results indicate that the values of λ may have a relationship with the rock strength.

For the Tornado Mountain rock fall site (see Section 2.2.1), the dimensions of selected rock fragments broken off the main rock fall in the lower part of the slope were recorded, together with the corresponding impact number. For rock fall A, where the final volume was about 1.42 cu. m (66 cu. yd) and mass = 3,750 kg (8,300 lb), it was estimated that fragments with a total volume of about 8 cu. m (10.5 cu. yd) broke off the body after impact #A10, of which a total of 10 blocks with a total volume of about 2 cu. m (2.6 cu. yd) broke off between impacts #26 and #31. It was not possible to distinguish rock fragments from above impact #A10, but it is expected that much of the loss of volume occurred early in the fall where the rock fell 40 m (130 ft) on to a bare rock bench. Using this data, the following approximate relationship was developed between the volume $\Omega_{(n)}$ at impact number *n*, the initial volume $\Omega_{(0)}$:

$$\frac{\Omega_{(n)}}{\Omega_{(0)}} = n^{-0.8} \tag{6.16}$$

Equation (6.16) shows that if $\Omega_{(46)} = 1.42$ cu. m (2 cu. yd) after 46 impacts, then the original volume $\Omega_{(0)}$ was about 30 cu. m (140 cu. yd); a rock fall with this volume is feasible for this strong, massive limestone. That is, about 95% of the original rock volume was lost during this fall over a total horizontal distance of about 610 m (2,000 ft).

In order to compare the results of Equations (6.15) for the Italian data and (6.16) for the Tornado Mountain events, the positions of the impacts at Tornado were converted into horizontal distances from the source *x*, with the fall stopping at a distance of *x* = 610 m (2,000 ft), and the ratio $\Omega/\Omega_0 = 1.42/30 = 0.047$. Substitution of these values into Equation (6.15) gives a value for the reduction coefficient λ, of 0.03. It is considered that this value for λ is consistent with the Italian data considering the relatively limited information available at Tornado Mountain.

The loss of volume (and mass) during the course of rock falls obviously has an effect on the potential, kinetic, and rotational energy losses. While the energy partition diagram in Figure 6.5 shows the loss of energy for each impact at the Ehime test site for a constant mass, Figure 6.7 shows the loss of energy for the mass decreasing at each impact point according to the relationship given in Equation (6.15).

The test block was a cubic-shaped concrete body with side dimensions of 0.6 m (2 ft), and an initial volume, Ω_0 = 0.22 m³ (0.30 yd³) (mass = 520 kg) (1,150 lb), and the total horizontal (x) distance traveled was 57 m or 190 ft (Figure 2.6). Assuming a reduction factor of λ = 0.02, the decrease in volume is calculated using Equation (6.15), for a final volume of 0.1 m³ (0.1 yd)³.

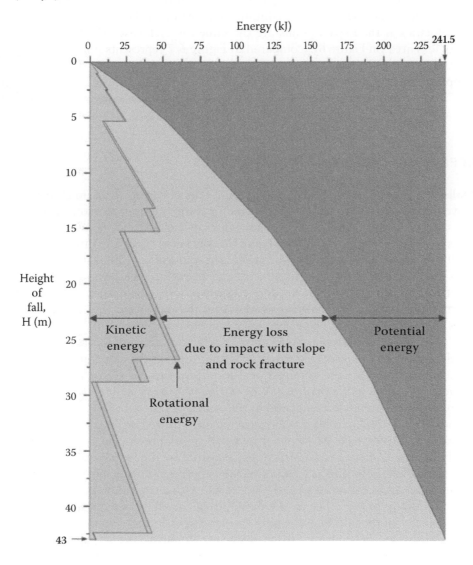

Figure 6.7 Energy partition plot for diminishing mass at the Ehime rock fall test site—cubic concrete block with initial side length 0.6 m (2 ft).

Figure 6.7 shows details of the changes in potential, kinetic, and rotational energy for the fall at the Ehime test site in Japan described in Section 2.2 for a mass that reduces with each impact according to Equation (6.15), from an initial volume of 0.22 m³ (0.3 yd³) to a final volume of 0.1 m³ (0.13 yd³). In preparing the plot shown in Figure 6.7, the calculated progressively reduced mass at each impact during the fall was used to calculate the corresponding reduced values of the potential and kinetic energies, as well as the reduced moment of inertia of the body to calculate the changes in rotational energy (see Equation [6.10]). For a cubic body with side length L, the moment of inertia, I is given by (see Table 4.1):

$$I = m\frac{L^2}{6}$$

Comparison of the two sets of energy partition lines shown in Figures 6.5 and 6.7 clearly demonstrate the effect of the rock mass diminishing during the fall, assuming that the translational velocities at the impact points are the same for each case. The increased width of the energy loss area on Figure 6.5 compared to Figure 6.7 represents the energy required to fracture the falling rock.

An application of the concept of rock mass loss during falls is in the design of rock fall fences. For a fence located at some distance from the source, the impact energy will be less than for a fence located close to the source.

6.6 EFFECT OF TREES ON ENERGY LOSSES

Rock falls often occur on forested slopes because falls occur in wet, cold climates that are conducive to tree growth, and falls are triggered by water and ice pressures. Also, growth of tree roots in cracks can be triggering mechanism as discussed in Section 1.4.

Studies have been carried out in both Japan (Masuya et al., 2009; Ushiro et al., 2006) and Europe (Dorren and Berger, 2012; Dorren, Berger, and Putters, 2006; Dorren and Berger, 2005) to quantify the effect of trees on rock fall behavior. The Japanese tests by Ushiro were part of the Ehime study (see Section 2.1) where fall velocities and run-out distances were measured for tests conducted before and after the trees were removed. At Ehime, the talus was sparely forested with pine and oak having trunk diameters in the range of 100 to 500 mm (4 to 20 in.). Measurements of impact velocities showed no significant difference in the velocities between tests conducted with the trees on the slope and after they were removed. Similar results were obtained at Tornado Mountain where the impacted trees were about 200 mm (8 in.) in diameter. Where trees were impacted by rock falls, they were sheared off with apparently no reduction in the velocity of the fall; Figure 2.7 shows a typical tree on Tornado Mountain that was impacted and sheared off by a rock fall.

Masuya et al. (2009) have developed a rock fall simulation model that incorporates the tree height, trunk diameter, and a probability density function expressed as the number of trees per square meter of slope surface. Collisions between a rock and a tree are modeled in the same manner as collisions with the slope, but with a restitution coefficient of 0.1 and a friction coefficient of 0.03. The model showed that the effect of trees is to approximately halve the velocity of the falls and to cause a wider dispersion of the falls on the slope compared to a bare slope. Section 3.5.3 discusses dispersion areas of rock falls on talus slopes.

In the European study by Dorren and Berger (2005), tests were conducted on two similar parts of a talus cone with a slope angle of 38° and a slope length of 302 m (990 ft). One part of the talus was denuded of trees, while the other part was forested with trees having an average diameter of 310 mm (12 in.) and density of 290 trees per hectare. This slope configuration allowed two near identical rock fall tests to be conducted—one slope with trees and one without trees. For each test, 100 rocks were rolled with their impacts and trajectories recorded at 25 frames per second by five high-speed cameras mounted in nearby trees; the rocks had an average diameter of 0.91 m (3 ft) and volume of 0.49 cu. m (0.6 cu. yd). The effect of the trees on the rock fall behavior was to reduce the average velocity from 13.4 m · s⁻¹ (44 ft · s⁻¹) to 11.2 m · s⁻¹ (37 ft · s⁻¹), and the maximum velocity from 14.8 m · s⁻¹ (49 ft · s⁻¹) to 11.6 m · s⁻¹ (38 ft · s⁻¹). From these results, as well as other tests on the resistance of trees to impact, it was possible to find a relationship between the diameter of tree trunks (measured at chest height, approximately where impacts occur), and the energy dissipated per tree for six tree species as shown in Figure 6.8. The tests also showed the relative wider dispersion of the rock falls by the tree impacts compared to bare slopes, similarly to that modeled by Masuya (2009).

The energy dissipation data shown on Figure 6.8 can be used to evaluate the results obtained from the Ehime test site and the Tornado Mountain rock fall events. In both these cases, the tree diameters were in the range of 100 to 200 mm (4 to 8 in.) where the energy dissipation would be negligible. This low-energy dissipation is consistent with the observation that the tree impacts at Ehime and Tornado Mountain had no significant effect on rock fall velocities and run-out distance.

Another characteristic of tree impact observed by Dorren and Berger is for rocks that are entirely or partially stopped by a tree. In some of these cases, the "whipping" action of the tree during impact can break off the upper part of the tree well above the impact point. Figure 6.9 shows a cedar tree with a diameter of 1.1 m (3.6 ft) that stopped a rock with a volume of about 5.5 cu. m (190 cu. yd) moving at an estimated velocity of 5 to 10 m · s⁻¹ (15 to 30 ft · s⁻¹); the approximate impact kinetic energy was 200 to 800 kJ (80 to 300 ft tonf). The tree was tilted and broke off at a height of about 14 m (46 ft) above the ground. Comparison of these impact parameters with Figure 6.8 confirms the capacity of this 1.1 m diameter tree to absorb this impact energy.

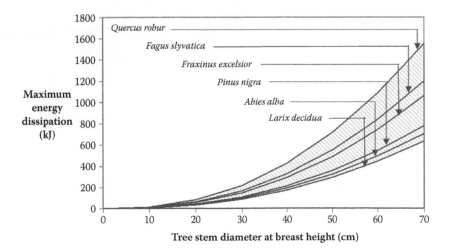

Figure 6.8 Relationship between maximum energy that can be dissipated by six different tree species, and the tree diameter, measured at chest height. (From Dorren, L. K. A. and Berger, F. (2005). *Tree Physiol*, 26, 63–71.)

Figure 6.9 Impact of a 5 cu. m (6.5 cu. yd) rock fall with kinetic energy of about 200 to 800 kJ (75 to 300 ft tonf) with a 1.1 m (3.6 ft) diameter cedar tree. Rock was stopped and upper part of tree was broken off about 14 m above base (Vancouver Island, near Ucluelet).

Chapter 7

Rock Fall Modeling

Design of rock fall protection structures requires information on the mass and impact velocity of the fall to determine the impact energy, and the fall trajectories to determine the height and location of the structure on the slope. These design parameters are usually obtained from computer simulation programs. Numerous modeling programs have been developed since the late 1980s that included commercial programs, and a variety of university research tools. The complexity of these programs varies from simple two-dimensional, lumped mass models to three-dimensional models in which the shape and size of the body can be defined and its orientation is tracked during the fall.

This chapter discusses, first, the general principles of rock fall modeling, the required input parameters, and the outputs that are generated. Second, analysis results are presented that model the rock falls for the five case studies described in Chapter 2; the modeling was carried out using the *RocScience* program *RocFall 4.0*. The purpose of carrying out these back analyses was to determine the values of the input parameters that are required to closely model the actual events. It is intended that these analyses will provide benchmarks for rock fall modeling because comparisons of the results of rock fall analysis programs with commonly used parameters show that calculated trajectories are often higher than those that occur in the field.

The consequence of unrealistically high, calculated trajectories is that many fences and barriers are higher, and more expensive, than required. For example, the author has observed fences that were designed using commercial software, where all the impacts were in the lower one-third of the structure.

Another purpose of providing the case study information in Chapter 2 is to provide actual rock fall data that can be used by developers of modeling software to calibrate their programs.

The modeling methods discussed in this chapter are based on the programs *RocFall 4.0* (*RocScience*, 2012) and **CRSP Colorado Rock Fall Simulation Program** (Pfeiffer and Higgins, 1995; Jones et al., 2000). These two programs are commercially available and the calculations methods are documented in detail. Also, the two programs can produce identical results with the use of appropriate input parameters. Figure 7.1 shows the results of a typical rock fall simulation for the Mt. Stephen site using *RocFall 4.0* (Section 2.1.1).

7.1 SPREADSHEET CALCULATIONS

A detailed analysis of rock fall trajectories and impacts for a lumped mass can be carried out on a spreadsheet using the basic principles of trajectories and impact mechanics described in the previous chapters. The calculation method is summarized in this section.

Chapter 2 provides documentation on rock fall events at five locations for a variety of slope geometries and geologic conditions. The information available for these events are

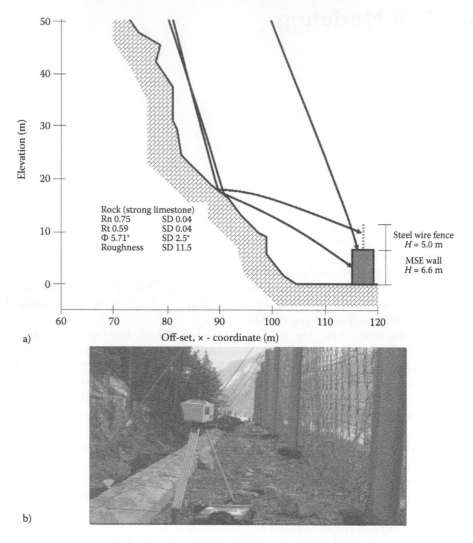

a)

b)

Figure 7.1 Rock falls at Mt. Stephen (see Section 2.1). (a) Simulation of rock falls showing three typical rock fall paths; (b) accumulation of falls that impacted fence on top of barrier.

the slope geometry and the coordinates of successive impact points. This information can be used to calculate the trajectory path, and the restitution and impact velocities of the body using Equation (3.4) that relates the $[x, z]$ coordinates along the path to the velocity (V_0), gravity acceleration g, and the restitution angle of the body relative to the x-axis, α. Measured values for the rebound angle θ_0 relative to the slope $[\theta_0 = (\alpha + \psi_s)$ where ψ_s is the slope angle] are available from the test site at Ehime in Japan (Section 2.1.3) and at Tornado Mountain in Canada (Section 2.2.1); values for θ_f are plotted in Figures 3.9(a,b). These two plots show both the range and the most frequent values of θ_0 that occur on natural slopes.

Values for the angle θ_0 at the Ehime test site were measured from the trajectory paths using accelerometers embedded in the test blocks, while at Tornado Mountain, it was possible to map impact points and locate the positions of 21 trees that had been impacted by rock falls, and measure the height of the impacts. If three points on the trajectory path are known (i.e., successive impact points with the slope, and an intermediate tree impact), then it is possible to exactly calculate the rock fall path and determine the values for α and θ_0.

Once successive trajectories have been calculated, it is possible to determine the impact and restitution velocities at each impact point, and their normal and tangential, and horizontal and vertical, velocity components. This information in turn allows the normal and tangential restitution coefficients to be calculated from the changes in velocity components that occur during impact. These same changes in velocity during impact allow the energy losses to be calculated at the impact points, as well as the energy gained during the trajectory due to gravitational acceleration. The result of these energy calculations is an energy partition diagram such as that shown in Figure 6.5.

A spreadsheet calculating these velocities, restitution coefficients, and energy changes can extend to 60 columns. This is an interesting amount of information that can be generated from the coordinates of successive impact points.

7.2 TERRAIN MODEL—TWO-DIMENSIONAL VERSUS THREE-DIMENSIONAL ANALYSIS

Early rock fall modeling programs were two dimensional (Piteau, 1980; Pfeiffer and Bowen, 1989), but with the development of methods of scanning rock slopes, such as Lidar, to produce digital terrain models (DTM) it has become possible to import DTM's into rock fall modeling programs and run three-dimensional analyses.

Modeling of a slope in two dimensions will only provide reliable results if the cross section is reasonably uniform along the slope. However, as discussed in Section 3.5.1, it is found that rock falls often accumulate in gullies in a similar manner to which water flows downslopes. Figure 3.12 shows a slope in which all the rock falls over a slope length of several hundred meters along the crest are concentrated in three narrow gullies. Modeling of rock falls on this slope using a two-dimensional program would give erroneous results because the model would not simulate the sinuous shape of the gullies and the generally uniform slope profile along the path of the gullies. A two-dimensional analysis of the slope in Figure 3.12 would show a series of ridges and hollows that would not be realistic model of the rock fall path.

7.3 MODELING METHODS—LUMPED MASS

The primary components of rock fall modeling programs are first, algorithms to calculate the rock fall impacts and trajectories and, second, routines to process the graphics of the slope and the rock fall paths. Of these components, the calculation of the impacts and trajectories can be readily accomplished if a lumped mass model is used, as is the case with the programs *RocFall 4.0* and *CRSP*; the calculations can be handled on a spreadsheet as described in Section 7.1. However, the calculations are more complex if the shape and size of the body are defined such that the program tracks the orientation of the body during both the trajectory and impact phases of the fall. Still more complexity is introduced if the rock breaks up during impact.

This section reviews the methods used to calculate trajectories and impacts for a lumped mass model such as used in the programs *RocFall 4.0* and *CRSP*. Section 7.4 discusses models that use bodies with defined dimensions and shapes.

The component of the modeling programs that processes the graphics is beyond the scope of this book.

7.3.1 Rock Fall Mass and Dimensions

The lumped mass model that is used in the programs *RocFall 4.0* and *CRSP* assumes that the body is infinitely small, which allows the use of Newtonian mechanics as described in the earlier chapters, to calculate impact and trajectory behavior, ignoring the effect of air friction. The mass of the body is used to determine energies from the calculated velocities, and it is assumed that the mass is constant during the fall; loss of mass during the course of a fall is discussed in Section 6.5.

Calculation of the effects of slope roughness (see Section 7.3.6) and rotational velocity (see Section 7.3.7) requires a value for the dimensions of the body. In *CRSP*, the default body shape is a sphere because it yields the maximum volume for a given radius, and the program calculates the radius of the body from the mass and rock density. It is also possible to define cylindrical and discoid shaped bodies in order to calculate corresponding moments of inertia.

7.3.2 Slope-Definition Parameters

Development of a rock fall model requires definition of the slope parameters that comprise the slope geometry, and the material properties. The slope geometry is defined, for a two-dimensional model, by a series of $[x - z]$ coordinates that join straight line segments. Each segment is assigned a slope material, the properties of which are the normal and tangential coefficients of restitution, the surface roughness and the friction angle. These parameters are discussed in the following sections.

Some models use a "spline" function to create a smooth curved surface between the defined points, but this approach does not allow a series of horizontal benches and steep cut slopes to be modeled.

7.3.3 Rock Fall Seeder

The rock fall models incorporate a "seeder" that defines the rock fall conditions at the origin of the falls in the model. In *RocFall 4.0*, the seeder parameters are the horizontal and vertical translational velocity components, the rotational velocity and the mass of the fall. These parameters allow falls to be modeled either where the source of the falls is the same point as the seeder ($v_x = v_z \approx 0$), or where the source of the falls is outside the model area and the falls have finite velocities at the origin of the model (v_x, $v_z > 0$). The relative values of v_x and v_z are selected to obtain the required trajectory angle.

With respect to angular velocity, either a negative or positive value can be input. However, it appears that the direction of rotation makes little difference to the calculated trajectories. Typical *RocFall 4.0* analyses show that the effect of rotational velocity is to flatten trajectories, which is consistent with impact mechanics theory for negative direction of rotation (see Section 4.6.4 and Figure 4.9).

7.3.4 Normal Coefficient of Restitution

The coefficient of restitution, e, relates the velocity of the falling body at the end of impact with the velocity at the moment of impact (see Chapter 4). It is also usual to examine the velocity changes in the normal (N) and tangential (T) components of velocity during impact and the corresponding normal e_N, and tangential e_T, coefficients of restitution. This section discusses the normal coefficient, while Section 7.3.5 discusses the tangential coefficient and its relationship to the frictional properties of the slope surface.

The normal coefficient of restitution is defined as follows:

$$e_N = -\frac{v_{fN}}{v_{iN}} \tag{4.6}$$

where v_{fN} is the normal component of the velocity at the completion of the impact $(t = f)$, and v_{iN} is the normal component of velocity at the moment of impact $(t = i)$.

It is recognized in the **CRSP** and **RocFall 4.0** models that e_N is not a material property but is dependent on impact conditions and to a lesser degree on the type of slope material. That is, for a high-velocity impact the body will tend to penetrate further into the ground in an inelastic impact, compared to a lower-velocity impact that will tend to be a more elastic and have a greater rebound height. This behavior is modeled by the following empirical equation that scales the normal component of the restitution velocity v_{fN}, according to the following equation:

$$v_{fN} = \left(\frac{v_{iN} \cdot e_N}{1 + \left(\dfrac{v_{iN}}{K}\right)^2} \right) \tag{7.1}$$

where v_{iN} is the normal component of the impact velocity, and K is a constant. The term $[(1+(v_{iN}/K)^2)^{-1}]$ is a scaling factor that is equal to 0.5 for the default value of $K = 9.14$ m $\cdot$ s^{-1} (30 ft $\cdot$ s^{-1}). Figure 7.2 is a plot of the scaling factor given by Equation (7.1) for values of K ranging from 9 to 40 m $\cdot$ s^{-1} (30 to 130 ft $\cdot$ s^{-1}). The curves show that v_{fN} equals v_{iN} (i.e., $e_N = 1$) for low-velocity impacts when the scaling factor approaches 1 for elastic impact.

The form of the relationship given by Equation (7.1) and plotted in Figure 7.2 is similar to that discussed in Section 5.2.1 and in Figure 5.5 that relates the normal coefficient of restitution e_N to the impact angle θ_i. That is, high normal velocities occur when the impact is close to normal, or $\theta_i > \sim 60°$. Under these conditions, e_N has a low value close to the basic coefficient of restitution determined by a drop test (see Figure 5.6). However, for shallow impacts where $\theta_i < \sim 30°$ and the normal velocity component is small, the value of e_N becomes larger.

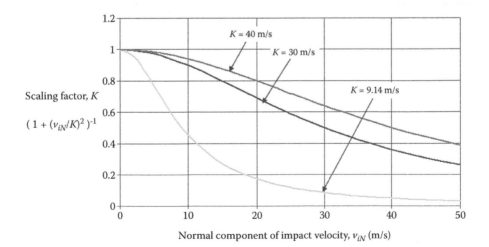

Figure 7.2 Plot of normal component of impact velocity and scaled normal coefficient of restitution from Equation (7.1).

Equation (7.1) shows that the maximum value of e_N scaled is 1.0, while the field data and impact mechanics show that e_N can have values greater than 1 for shallow impact angles. The relationship between θ_i and e_N is given by the equation:

$$e_N = 19.5 \, \theta_i^{-1.03} \tag{5.4}$$

7.3.5 Tangential Coefficient of Restitution and Friction

The tangential coefficient of restitution e_T, is defined by the ratio of the final and initial tangential velocities during impact as follows:

$$e_T = \frac{v_{fT}}{v_{iT}} \tag{4.7}$$

Tangential velocity changes depend on the extent of slip and rolling that occurs during impact and the effects of friction at the rock-slope contact.

In **CRSP**, the tangential coefficient of restitution is scaled by a friction factor $f(F)$, and a scaling factor SF that incorporate the impact translational and rotational velocities and the normal coefficient of restitution (Pfieffer and Bowen, 1989). In **RocFall 4.0**, analysis values for e_T incorporate data from back analysis of rock falls.

Since e_T is related to an effective friction coefficient μ', guidance on appropriate values for e_T can be obtained from measurements of fall velocity related to the fall height and the slope angle (Japan Road Association, 2000); Table 3.1 lists values for μ' for characteristics of different slope materials. For example, $\mu' \approx 0.05$ for "smooth, strong rock surfaces with no tree cover," and $\mu' \approx 0.35$ for "talus with angular boulders exposed at the surface, no tree cover." In addition, values for e_T have been calculated by back analysis for the five case studies discussed in Chapter 2 and are listed in Table 2.1; these values have been used in the **RocFall 4.0** modeling of these five sites as presented in Section 7.5.

In *RocFall 4.0*, the friction angle φ input parameter is related to calculate distance that the body slides down the slope when the trajectory phase of the fall is complete. The value of φ used in the analysis is a function of the body shape since a spherical body will have less frictional resistance to sliding than a tabular body. The distance L that a body will slide along a surface dipping at ψ_s degrees and with a friction angle φ is

$$L = \frac{(v_{iT}^2 - v_{fT}^2)}{2 \cdot g(\sin\psi_s - \cos\psi_s \cdot \tan\varphi)} \tag{7.3}$$

where v_{iT} is the tangential velocity at the start of sliding and v_{fT} is the velocity when the body has slid distance L (Stevens, 1998).

7.3.6 Surface Roughness

The surface roughness of the slope materials has a significant effect on rock fall behavior, that is, trajectory heights generally increase with increased roughness.

Surface roughness can be quantified by relating the perpendicular variation s of the slope from the average slope inclination, within a slope distance equal to the radius r of the body (Pfieffer and Bowen, 1989) as shown in Figure 7.3. The angle ε is defined by the dimensions r and s as follows:

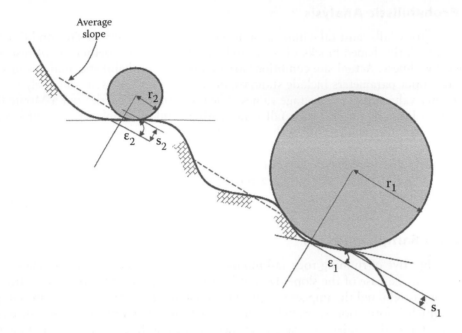

Figure 7.3 Relationship between slope roughness (ε) and radius of rock fall (r).

$$\varepsilon = \tan^{-1}\left(\frac{s}{r}\right) \qquad (7.4)$$

Figure 7.3 shows that the value of the angle ε diminishes with increasing radius of the body. That is, for approximately the same slope roughness, the roughness angle will diminish as the body size increases, ($r_1 > r_2$) and ($\varepsilon_2 > \varepsilon_1$).

In *CRSP*, the slope roughness is input as the dimension s from which the angle ε is calculated using the defined radius of the spherical body. In *RocFall 4.0*, the roughness is defined as a standard deviation in the dip angle of the slope segment.

7.3.7 Rotational Velocity

Rock falls rotate as the result of the moment generated by the tangential velocity and the frictional resistance at the contact of periphery of the body and the slope surface. Rotation of the body has two main effects on rock fall behavior. First, rotation causes the trajectories to be flatter than for a non-rotating body (see Section 4.6.4), and second, the rotating body has rotational energy that contributes to the total impact energy (see Section 6.2).

Body rotation is incorporated in the **CRSP** and *RocFall 4.0* models by calculating the rotational velocity ω for a body with radius r with tangential translational velocity v_T from the following relationship:

$$\omega = \frac{v_T}{r} \qquad (7.5)$$

The rotational energy (RE) of the body can be calculated using the moment of inertia, I, for the selected body shape and the calculated rotational velocity (RE = ½ $I \cdot \omega^2$).

7.3.8 Probabilistic Analysis

Modeling of rock falls must take into account the natural variability of site conditions that includes irregularly shaped blocks of rock, and variability of the coefficients of restitution and slope roughness. Actual site conditions are modeled by probabilistic analyses in which the average input parameters include standard deviations that represent the likely range of the parameter values from the average values. The program then carries out a Monte Carlo analysis for a large number of rock fall runs, with a random number generator selecting parameter values for each run from the probability distributions defined by the standard deviations. The result of the Monte Carlo analysis is a plot of all the analyzed rock falls that shows the likely distributions of rock fall behavior that may be expected.

Examples of probabilistic analyses are provided in Section 7.5 for the five case studies described in Chapter 2.

7.3.9 Data Sampling Points

The usual objective of running rock fall models is to design structures and ditches to protect facilities at the base of the slope. Design information required for these structures are the trajectory height and the impact energy. The modeling programs incorporate sampling points at which information is provided on distributions of analysis data—total, translational and rotational energies, velocity, and trajectory height. By moving the sampling point across the slope, it is possible to identify the location with the minimum energy and/or the lowest trajectory.

Examples of data generated at sampling points are provided in the analyses of the case studies in Section 7.5.

7.4 MODELING METHODS—DISCRETE ELEMENT MODEL (DEM)

Rock falls models that use a lumped mass in which the mass of the body is defined, but the mass is concentrated in a point are described in Section 7.3. Alternatively, discrete element models (DEM) can be used in which the mass, dimensions, and shape of the body are defined, and the body can break into smaller fragments as it impacts the slope during the fall (Zhang and Rock, 2012; Chen et al., 2013).

In DEM analysis, the body is made up of a collection of small spheres, in tight tetrahedral packing, connected with appropriate constitutive models to describe rigidity, heterogeneity, and fracture of the model. This model can accurately replicate trajectories, rolling, sliding, launching behavior of the body, and crack propagation within the body. With respect to impact of the body with the slope, the parameters related to impact that are required for DEM modeling are damping, stiffness, and friction coefficients.

As of 2013, commercial modeling programs in which the mass, shape, and size of the body are defined are not widely available.

7.5 MODELING RESULTS OF CASE STUDIES

For each of the five documented rock fall sites described in Chapter 2, a computer simulation has been run using the program **Rocfall 4.0** (*RocScience*, 2012). These analyses demonstrate the operation and results of the program and determine the site parameters that are required to produce calculated rock fall paths that closely follows the actual paths.

In defining the modeling parameters required to duplicate the actual field results, it was found that very fine adjustments were necessary in the average values of parameters. That is, analyses of active rock fall sites using apparently appropriate input parameters may show that falls stop partway down the slope, or that unrealistically high trajectories are generated.

7.5.1 Rock Fall Model of Mt. Stephen Events

Section 2.1.1 in Chapter 2 describes the rock fall conditions at Mt. Stephen in the Rocky Mountains near Field, British Columbia, in Canada. Figures 2.1 and 2.2 show, respectively, photographs of a portion of the site, and a typical section of the lower part of the slope. This is a highly active rock fall site, due to the topography, geology, and weather. That is, the mountain is about 2,000 m (6,550 ft) high at an overall face angle of about 50° so that rock falls can fall from great heights and attain high velocities, with most falls reaching the base of the slope. The geology comprises horizontally bedded limestone and shale where the limestone is much stronger than the shale and occurs in thicker beds; the limestone contains sets of vertical joints. The relatively rapid weathering of the shale compared to the limestone results in the formation of unstable overhangs and columns in the limestone that are the sources of the rock falls. The other factor causing the high rock fall frequency is the weather—very cold winters forming ice on the slope, and rainfall during the spring and fall. Figure 7.4 shows the lower 120 m (400 ft) portion of the slope.

The protection provided for these severe rock fall and snow avalanche conditions comprises a 6.6 m (21.5 ft) high MSE (mechanically stabilized earth) wall, with a 5 m (16.5 ft) high steel-wire fence along the top (Figure 7.1). This barrier has been very successful in protecting the railway from both types of hazard.

The actual rock falls at Mt. Stephen have been modeled using the program *RocFall 4.0*. Figure 7.4 shows the calculated trajectories for three falls, and Figures 7.5 and 7.6 show,

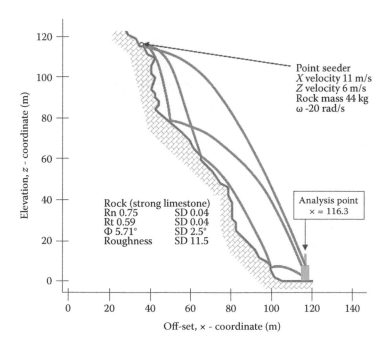

Figure 7.4 Simulation of rock falls at Mt. Stephen for three calculated rock trajectories.

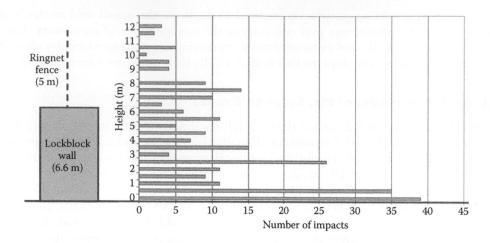

Figure 7.5 Calculated vertical distribution of impact points on barrier at Mt. Stephen.

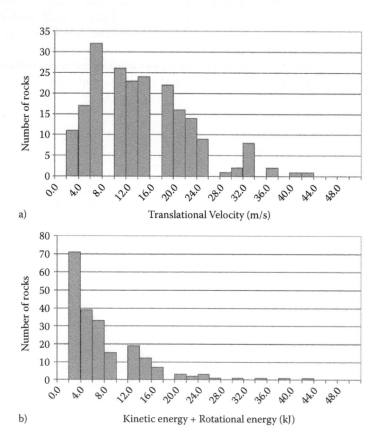

Figure 7.6 Analysis using *RocFall 4.0* of rock falls at Mt. Stephen at barrier, analysis point *x* = 116.3 m (a) translational velocity distribution, (b) total energy (KE + RE) distribution.

respectively, graphs of the distributions of impact location, and velocity and energy at the barrier location.

The material types used in the analysis was rock for the entire slope, with soil in the ditch behind the barrier. For this case study, all the rock falls were contained by the barrier so it was not possible to use the back analysis feature in *RocFall 4.0*. The parameters for the slope, including the standard deviations (SD) quantifying the range of the values, required to simulate the rock falls are shown in Figure 7.4. It is assumed that all rock falls originated higher on the slope than elevation 118 m (390 ft.) so the x and z seeder velocities have values that generate trajectories that are consistent with observed field conditions.

Observations of rock fall impacts on the MSE barrier and the fence provided reliable information on the impact locations. In total, it was possible to identify 466 impact points as either chips on the concrete blocks forming the face of the MSE wall, or as deformations of the steel wires in the fence. Analysis of the impact locations showed their vertical distribution, with impacts over the full 11.6 m (38 ft) height of the structure with the maximum impacts at the base and the number of impacts decreasing with height. The horizontal distribution of impacts showed that most occurred where the slope geometry included the lower face sloping at about 45° on which many blocks impacted and then generated a trajectory that impacted the barrier (Figure 7.4). Figure 7.5 shows the calculated vertical distribution of the impacts on the barrier which closely match the actual impact locations.

It was also possible to measure the dimensions of blocks that had impacted the fence and were then lying on top of the wall. As would be expected for this condition where rocks had fallen from a considerable height and impacted the slope several times, the maximum block size was only 300 to 500 mm (12 to 20 in.) approximately. Figure 7.1(b) shows the dimensions of the typical blocks accumulated on the top of the wall.

For the typical trajectories shown on Figure 7.4, the calculated velocities of falls that impacted the barrier were up to 44 m · s⁻¹ (145 ft · s⁻¹). For a block with a mass of 50 kg (110 lb) (ellipsoid with major axis: $2a = 2b = 0.4$ m, $2c = 0.2$ m, volume = 0.02 cu. m (0.03 cu. yd)), the impact energy would be about 60 kJ (22 ft tonf).

The calculated velocities and energies plotted in Figure 7.6 appear to be less than the actual values where falls originate on the steep slope from heights of hundreds of meters above the barrier and attain high velocities.

7.5.2 Rock Fall Model of Kreuger Quarry, Oregon, Test

The purpose of the rock fall tests carried out in the Krueger Quarry in Oregon was to determine the required dimensions and configurations of catchment areas to contain rock falls on highways (Pierson et al., 2001). The tests involved dropping rocks down an excavated rock face in to a catchment area at the base of the cut, and measuring both the first impact point and the farthest distance that the block rolled past the base of the cut (See Section 2.1.2). The cut heights ranged from 8 to 24 m (25 to 80 ft) and the face angles from vertical to 45°, with the slopes of the catchment area being horizontal, and sloped at 4H:1V and 6H:1V toward the cut face. The catchment areas were all uniform surfaces with no barriers or depressions since this is the configuration required for catchment areas ("recovery zones") on highways in the United States. In total, 11,250 separate rock fall tests were conducted.

Although no specific information on rock fall trajectories is available from high-speed camera images, for example, it is possible to determine likely trajectories that are mathematically feasible, from the slope and catchment geometry, and the records of the first impact points. Figure 7.7 shows calculated trajectories for a fall on a 15 m (50 ft) high cut with a face angle of 76° (1/4H:1V) where the first impact point at $x = 11.8$ m (40 ft) is the 95 percentile for this slope configuration. That is, 95% of all rock falls first impacted the ditch

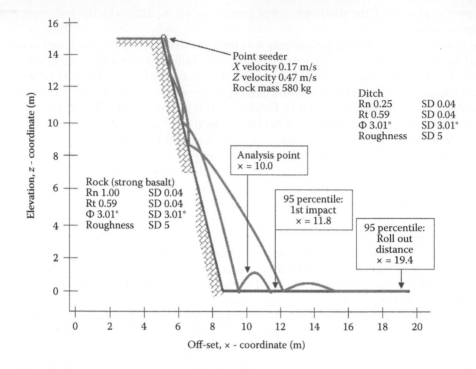

Figure 7.7 Calculated trajectories for two 580 kg (1,280 lb) rocks at Krueger Quarry rock fall tests on 15 m (50 ft) high cut at a face angle of 76°; refer to Figure 10.3 for first impact and roll out distance.

closer to the slope than the trajectory shown. Figure 7.7 also shows the input parameters for *RocFall 4.0* that are applicable to slope configuration. The point seeder in this case has low values for the *x* and *z* velocities because the rocks were pushed off the crest of the cut.

Figure 7.7 also shows an analysis point located at $x = 10$ m (33 ft), or about 1.4 m (5 ft) from the base of the cut; this location represents the 70th percentile of the first impact points. The calculated distributions of translational velocities and energies at $x = 10$ m (33 ft) are shown on Figure 7.8.

7.5.3 Rock Fall Model of Ehime, Japan, Test Site

In 2003 a series of rock fall tests were carried out at a test site in Ehime Prefecture on Shikoku Island in Japan, as part of an extensive testing program that started in about 1961, located at least 16 other sites around Japan (see Sections 2.1.3 and 2.2.1). The Ehime tests were comprehensive in terms of the number of block shapes tested and the range of site conditions. The tests were documented in detail using 14 high-speed cameras on the slope, and accelerometers sampling at 2 kHz embedded in concrete blocks. The data collected allowed the precise positions of impact points and trajectory paths to be determined throughout each fall, as well as the translational and rotational velocities.

The test slope was 42 m (140 ft) high natural slope, comprising a 26 m (85 ft) high rock face at an angle of 44° in a horizontally bedded sandstone and shale, with a 16 m (50 ft) high talus slope forming the lower part of the slope. The talus cone was sparsely vegetated with pine and oak trees with trunk diameters of 100 to 200 mm (4 to 8 in.).

The test conditions were as follows. The test bodies comprised concrete cubes, side length 0.6 m (2 ft) and weight of 520 kg (1,150 lb), concrete spheres, diameter 0.54 m (1.8 ft) and

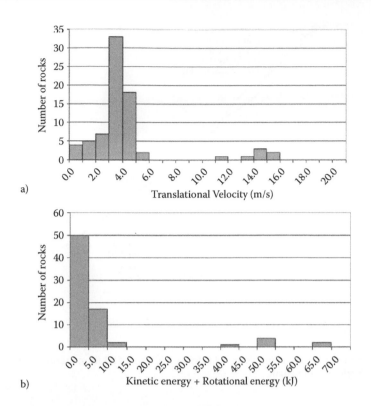

Figure 7.8 Analysis using *RocFall 4.0* of rock falls at Krueger Quarry for 15 m (50 ft) high cut at a face angle of 76°, analysis point *x* = 10 m (33 ft); (a) translational velocity distribution; (b) total energy (KE + RE) distribution.

weight of 200 kg (440 lb), and blocks of rock with masses ranging from 120 kg to 2060 kg (260 to 4550 lb). The tests involved rolling 10 cubes, 10 spheres, and 20 blocks of rock, with half the tests being run on the natural treed slope, and the second half after the trees had been removed.

The collected data were used to determine the fall paths and trajectory heights, as well as the translational and rotational velocities. A photograph of the slope is shown in Figure 2.5, and details of the interpreted data are presented in Figure 3.5—trajectory heights normal to the slope; Figure 3.6—translational velocities; Figure 3.9—distribution of restitution angles; and Figure 3.10—angular velocities.

Figure 7.9 shows a simulation, using *RocFall 4.0*, of the actual fall path of a concrete cube as shown on Figure 2.6. The input parameters for the rock and talus material properties and the seeder values are also shown on Figure 7.9.

Figure 7.10 shows a comparison between the actual and calculated trajectory heights for a test with a concrete cube. The two sets of data are reasonably consistent except for the talus impact at *x* = 34 m where the calculated trajectory is 50% higher than the actual trajectory. Figure 3.5 shows the measured trajectory heights for all the tests, where 99% of the trajectory heights were less than 2 m (7 ft), and 95% were less than 1.5 m (5 ft).

Figure 7.11 shows the distributions of the velocity and total energy of 24 tests of concrete cubes. The analysis point (*x* = 52.1 m) is located at the base of the talus slope where a rock fall protection structure may be located. The calculated velocities at the analysis point can be compared with the actual velocities at this location shown on Figure 3.6 where *H* = 41 m

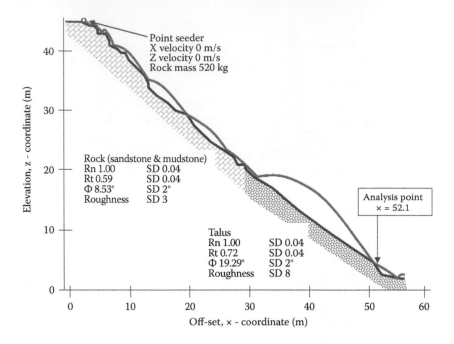

Figure 7.9 Calculated trajectory using *RocFall 4.0* for a 520 kg (1,150 lb) concrete cube at the test site in Ehime Prefecture in Japan.

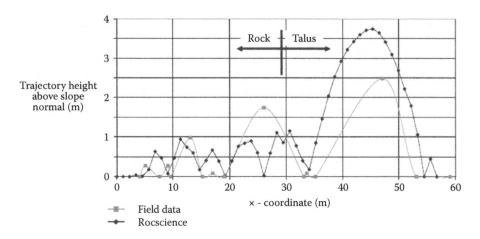

Figure 7.10 Trajectory height envelope comparison between field results and *RocFall 4.0* simulated results for concrete cube at test site in Ehime Prefecture in Japan.

in Figure 2.6, equivalent to $x = 52.1$ m in Figure 7.9, the actual range of velocities is 5 to 15 m · s⁻¹.

7.5.4 Rock Fall Model of Tornado Mountain Events

Tornado Mountain is located in southeast British Columbia near the town of Fernie (see Section 2.2.2). The site of the rock falls comprises a 50 m (165 ft) high, near vertical rock face in very strong, blocky limestone above a talus/colluvium slope; the talus in the upper

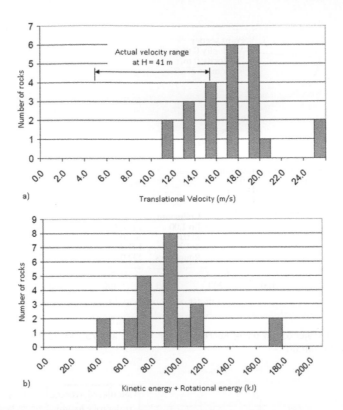

Figure 7.11 Analysis using *RocFall 4.0* of rock falls at Ehime test side for 42 m (140 ft) high natural slope comprising bedrock (26 m) (85 ft) and talus (16 m) (50 ft) at analysis point x = 52.1 m; (a) translational velocity distribution; (b) total energy (KE + RE) distribution.

part of the slope is at an angle of about 35°, while the colluvium forming most of the slope is at an angle of about 22° (see Figure 7.12, and Figures 2.8 and 2.9). The colluvium forming the lower slope is a mixture of gravel and soil forming a uniform slope with no significant irregularities; the slope is sparsely vegetated with trees having diameters ranging from about 300 to 500 mm (12 to 20 in.).

In 2004, two separate rock falls, with masses of A = 3,750 kg (8,300 lb) and B = 5,600 kg (12,400 lb), occurred from a source area on the limestone cliff. The rocks traveled total distances of 740 m (2,450 ft.) down the slope before impacting a horizontal bench where most of the kinetic energy was absorbed, and the rocks stopped within 30 m (100 ft). Because the two rocks followed slightly different paths and no previous falls had occurred in this area, it was possible to identify and map most of the impact points in the slope, including 21 trees that were sheared off by the falls. This information on the impact locations allowed velocities and trajectories to be calculated as discussed in Section 2.2.2.

Figure 7.12 shows the results of the *RocFall 4.0* simulation of Tornado Mountain fall A, with the input parameters required to produce trajectories that reasonably closely match the field measurements. Figure 7.13 compares the calculated trajectory heights with the average height measured in the field of 1.5 m (5 ft.) and shows that most of the calculated trajectories are significantly higher than actual heights. This discrepancy between calculated and actual heights can lead to the construction of protection structures that are higher than required. It is the author's experience that calculated trajectories are often higher than actual trajectories.

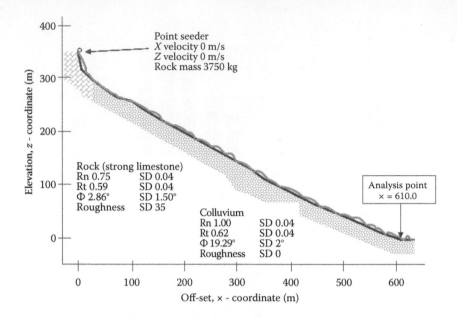

Figure 7.12 Calculated trajectories using *RocFall 4.0* for fall A, a 3,750 kg (8,300 lb) limestone block at Tornado Mountain.

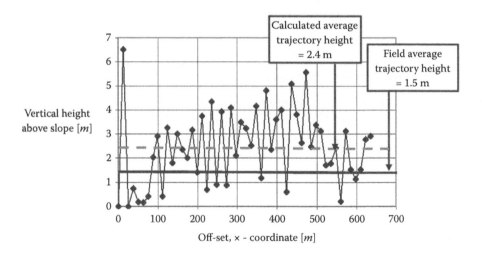

Figure 7.13 Trajectory height envelope from *RocFall 4.0* simulated results for a 3,750 kg (8,300 lb) limestone block at Tornado Mountain.

Figure 7.14 shows the calculated distributions for the velocity and energy at an analysis point located just above the bench on which the railway is sited, and where a fence or barrier would likely be constructed. The actual impact properties of fall A at this location were a velocity of 18 m · s^{-1} (60 ft · s^{-1}) and a kinetic energy of 600 kJ (220 ft tons) as shown in Figure 9.1. If the block has a moment of inertia of 800 kg · m^2 and is rotating at 15 rad · s^{-1}, then the rotational energy, RE = ($\frac{1}{2}$ I · ω^2) = 90 kJ, and the total energy is 690 kJ (255 ft tonf). These actual impact velocity and energy values are plotted on Figure 7.14, showing that they are at the low end of the calculated distributions.

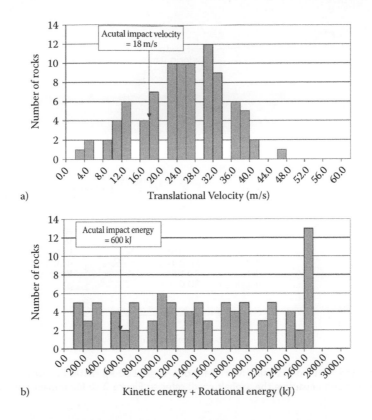

a)   Translational Velocity (m/s)

b)   Kinetic energy + Rotational energy (kJ)

Figure 7.14 Analysis using *RocFall 4.0* of rock falls at Tornado Mountain, block A at analysis point *x* = 610 m (a) translational velocity distribution; (b) total energy (KE + RE) distribution.

The comparatively high calculated velocities may account for the calculated trajectories being higher than the actual trajectories.

7.5.5 Rock Fall Model of Asphalt Impact Event

A single rock fall occurred from the crest of a natural 138 m (450 ft) high slope made up of a 56 m (180 ft) high rock slope at an angle of 60°, a 70 m (230 ft) high colluvium slope at an angle of 42°, and a 10 m (33 ft) high rock cut above a highway (see Section 2.3). Figure 7.15 shows two calculated trajectories for the full fall height, together with the *RocFall 4.0* parameters required to generate this simulation.

Figure 7.16 shows the calculated distributions of impact velocity and energy just before the impact with the asphalt at the analysis point *x* = 140 m. These calculated values can be compared with the precise trajectory of the fall from the crest of the rock cut to just before and after impact with the asphalt as shown in Figure 2.10 and discussed in Section 2.3. For a block with a mass of 500 kg (1,100 lb), a moment of inertia, $I = (m \cdot k^2) = (500 \cdot 0.295^2) = 43.5$ kg $\cdot$ m², an impact velocity of 21.5 m $\cdot$ s⁻¹ (70 ft $\cdot$ s⁻¹), and a rotational velocity of 15 rad $\cdot$ s⁻¹, the impact kinetic and rotational energies are 115 kJ (42.5 ft tonf) and 4.8 kJ (1.8 ft tonf), respectively. These actual velocity and energy values are plotted on the calculated distributions.

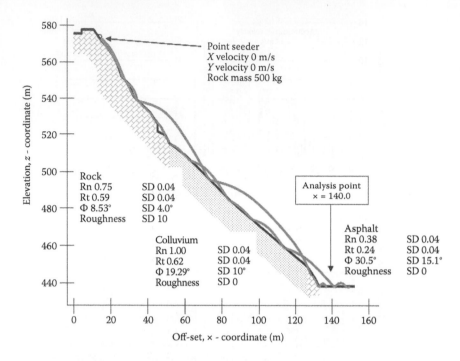

Figure 7.15 Calculated trajectories using *RocFall 4.0* of a single 500 kg (1,100 lb) rock fall from the crest of the slope and impacting the asphalt road; refer to Figure 2.10 for impact details on asphalt.

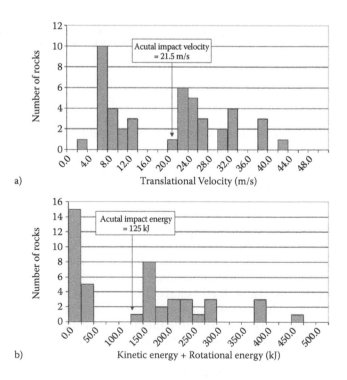

Figure 7.16 Analysis using *RocFall 4.0* of rock fall impacting asphalt at analysis point *x* = 140 m; (a) translational velocity distribution; (b) total energy (KE + RE) distribution.

Table 7.1 Summary of input parameters used in *RocFall 4.0* to stimulate case study rock falls

Site no.	Rock fall site	Slope material	Normal coefficient of restitution, e_N mean/SD	Tangential coefficient of restitution, e_T mean/SD	Friction angle (deg) mean/SD	Slope roughness (deg) SD
1	Mt. Stephen, Canada	Rock	0.75/0.04	0.59/0.04	5.71/2.50	11.50
2	Oregon ditch study (rock face impact)	Rock	1.00/0.04	0.59/0.04	3.01/3.01	5.00
2	Oregon ditch study (ditch impact)	Rock	0.25/0.04	0.59/0.04	3.01/3.01	5.00
3	Ehime, Japan (rock slope)	Rock	1.00/0.04	0.59/0.04	8.53/2.00	3.00
3	Ehime, Japan (talus slope)	Talus	1.00/0.04	0.72/0.04	19.29/2.00	8.00
4	Tornado Mountain, Canada	Rock	0.75/0.04	0.59/0.04	2.86/1.50	35.00
4	Tornado Mountain, Canada	Colluvium	1.00/0.04	0.62/0.04	19.29/2.00	0.00
5	Highway	Asphalt	0.38/0.04	0.24/0.04	30.5/15.10	0.00
5	Highway	Rock	0.75/0.04	0.59/0.04	8.53/4.00	10.00
5	Highway	Colluvium	1.00/0.04	0.62/0.04	19.29/10.00	0.00

In RocScience 4.0, default SD for e_N and e_l = 0.04.

7.6 SUMMARY OF ROCK FALL SIMULATION RESULTS

Section 7.5 describes the results of rock fall simulations for the five case studies described in Chapter 2. The simulations show the calculated trajectory heights and the velocity and energy distributions at selected analysis points that correspond to locations where reliable field data are available. It was found that it is possible to simulate actual field conditions, although calculated trajectory heights and velocities tend to be higher than actual heights and velocities. Furthermore, it was found that the calculated results are very sensitive to the input parameters, in terms of both the selected values and their standard deviations. Table 7.1 lists the input parameters that were used in the simulations.

As a general comment on the simulation of rock falls, it is difficult to obtain calculated results that are close to actual conditions without first knowing the actual field conditions in order to calibrate the calculations. That is, small changes in the input parameters can produce large changes in the calculated results that may appear to be reasonable but are in fact incorrect. It is hoped that the simulation results presented in this chapter will be of assistance in producing reliable simulations.

Chapter 8

Selection of Protection Structures

The selection of rock fall protection structures that are appropriate for the site conditions depends on a combination of factors that include design impact energy, topography, slope geometry, and the type of facility that is to be protected. This chapter discusses rational methods to select protection structures suitable for site conditions based on the relationship between the return period for rock falls and their mass, and the application of risk management and decision analysis to match the level of protection with the consequences of a rock fall.

Figure 8.1 shows a rock fall that has broken through the roof of a reinforced concrete shed; this impact energy has clearly exceeded the design capacity of the shed.

8.1 IMPACT ENERGY—DETERMINISTIC AND PROBABILISTIC DESIGN VALUES

A primary design parameter required for rock fall protection structures is the impact energy that the structure is required to withstand. The design energy is the total of the kinetic and rotational energies, for which the components are the mass and shape of the rock fall, and the translational and rotational velocities; the rotational energy is usually a small portion (about 10% to 20%) of the total energy.

The design energy can be expressed using either deterministic or probabilistic methods, depending on site conditions. The following two examples illustrate typical project conditions and how they influence the selection of design energies.

Deterministic design energy example, single hazard location—a 10 km length of highway, with high traffic volumes and very low tolerance for traffic disruptions, is located in a steep-sided canyon and has a single rock fall hazard at a tunnel portal. It has been decided that a reinforced concrete rock shed, which will have a long design life and require little maintenance, is appropriate to provide highly reliable protection against rock falls. The structural designers for the shed require that the design energies be specific (deterministic) values that they can use to prepare a design that can withstand these impacts. In accordance with structural design procedures, the design energy is expressed as a service limit state and an ultimate limit state, as defined in Section 8.2 below.

Probabilistic design energy example, multiple hazard locations—a 25 km (15.5 miles) length of railway with high traffic volumes located in the same canyon as the highway discussed in the previous paragraph, has 18 rock fall hazard locations. At each of these hazard locations, mitigation measures have been implemented that include ditches, slide detector fences (see Section 10.3), and a number of wire-mesh fences. The mitigation measures provide a level of protection that is significantly better than no protection, but the railway accepts that an occasional, large rock fall will damage

Figure 8.1 Rock fall impact that exceeded ultimate design capacity of concrete shed.

the protection structures and that repairs will be necessary. It has been decided that this is an economical approach to rock fall protection because the cost is prohibitive to provide highly reliable protection such as concrete sheds at all 18 locations. Under these conditions, probabilistic methods, as described in Section 8.3 below, are used to determine the design energy that the structure is designed to withstand without damage, with the understanding that the occasional fall with energies greater than the design energy will cause damage and service disruptions. The cost of both damage and service disruptions would be considered in selecting the design energy appropriate for the site.

Using probabilistic design methods, the design would provide protection for 96%, for example, of all fall energies, but that damage to the structure will occur for 4% of the fall energies. Section 8.4 describes how the relationship between return periods and rock fall mass can be determined, with large falls occurring much less frequently than small falls. Complete (100%) protection would require relocating the railway in a tunnel, whereas protection for 96% of the falls can be achieved by installing rock fall fences. The return period/rock fall mass relationship would show if the substantially greater cost of driving a tunnel, compared to installing fences, is justified.

8.2 IMPACT ENERGY—SERVICE AND ULTIMATE STATES ENERGIES

The design of rock fall protection structures is analogous to the seismic design of bridges and buildings where the impact energy and the level of ground shaking that the structures must resist are uncertain. One method of addressing this uncertainty is to use Limit States Design (LSD) methods in which the service limit state (SLS) and ultimate limit state (ULS) loads are applied (Canadian Geotechnical Society, 2006).

For example, some seismic codes specify two levels of earthquake shaking: a Serviceability Level Earthquake (SLE) that occurs relatively frequently and a Maximum Credible Earthquake (MCE) that has a low probability of occurring during the design life of the structure. Structures are designed to resist the SLE with little or no damage so that they function for their intended use, and to resist the MCE without collapse. However, the MCE may cause extensive damage to the structure requiring significant repairs or even demolition.

In the design of rock fall protection structures, the two-limit states are the service and ultimate impact energies; in the European design code for rock fall protection structures (ETAG–27), it is specified that the ultimate maximum limit state energy (MEL) is three times the service limit state energy (SEL).

The service and ultimate limit state energies to which rock fall containment structures will be subjected may be described as follows (see Figure 8.2):

- **Service limit state**—The service limit state is the impact energy to which the structure will be commonly subjected, and which it can sustain without damage or need for maintenance. That is, the structure may be impacted hundreds of times by rocks with energies up to the service energy, and the only maintenance requirement is to remove accumulated rock falls when they reach a point that they are reducing the capacity of the structure (Figure 8.2(a)). The service energy will be determined by examining the usual rock fall dimensions from observations of the geology in the source area, and the dimensions of blocks in the run-out path of the site. The dimensions of the block that will impact the protection structure should also take into consideration how much the rock is likely to break into smaller fragments during the fall, depending on the type of surfaces it will impact along the fall path, that is, bare rock or soft soil (see Section 6.5). The velocity used in calculating the service energy would depend on the fall height, the slope angle of the run-out path, and the irregularity of the slope surfaces (see Section 3.2). The service energy would be calculated using the most common values for all these parameters.
- **Ultimate limit state**—The ultimate limit state is the energy that will cause damage to the structure without collapsing; the structure will require maintenance and repair but can then be put back into service. For example, in designing a reinforced concrete rock fall shed, the ultimate energy would cause cracking and spalling of a roof beam, but this would not impede passage of traffic once the debris had been removed (Figure 8.2(b)). It would be possible to schedule maintenance to replace or repair the beam at a convenient time for operation of the facility.

Ultimate service energy could be determined by one of two methods, or a combination of the two. First, examination of the site may clearly show, for example, that the rock in the source area contains distinct joint sets, and that the largest rock fall that may occur is defined by the spacing and persistence of these sets. The second method of defining the ultimate energy is to extrapolate the service energy design parameters to the largest dimensions and highest velocities that may be expected at the site. With respect to the dimensions of the largest rock fall, this may be greater than any existing fall at the site, so statistical methods, such as the Gutenberg–Richter cumulative annual frequency technique and the Gumbel extreme value theorem, would be used for the extrapolation (see Section 8.4). With respect to the highest velocity that may occur, this is usually limited by the characteristics of the fall path—the effective friction coefficient of the surface and its slope angle—as defined by Equation (3.13).

It would be usual to combine the site observation and statistical methods because a certain amount of judgment will still be required to identify the common rock fall dimension and extrapolate this to the largest fall that may occur.

Figure 8.2 Examples of service and limit states energies. (a) Fence with no damage containing rock falls with impact energies less than service limit energy; (b) Roof of rock shed with spalled concrete where impact energy just exceeded ultimate limit energy.

These statistical methods are discussed in Section 8.4, and have the objective of defining the relationship between the rock fall mass and its return period. This selected return period may be consistent with return periods for other design parameters, such as seismic ground accelerations and floods that are being used on the project.

8.3 IMPACT ENERGY—PROBABILITY CALCULATIONS

Probabilistic methods are a useful design tool for rock fall analysis because of the uncertainty in the velocity and mass of the rock falls; the probabilistic method quantifies this uncertainty to assist in the determination of realistic design energies. In comparison, a deterministic approach described in Section 8.1 would calculate the impact energy for falls with the largest mass considered feasible, falling with high velocity from the crest of the slope. While this would be about the largest energy that may impact the structure, it would be a rare event since falls can originate from any height on the face, and falls tend to break into smaller fragments as they impact the rock face during the fall.

The result of probabilistic calculations is a probability distribution of the impact energy showing the percent of rock fall energies that exceed a specified energy. As discussed in Section 8.1, the probabilistic design approach is applicable where the rock fall mitigation strategy is to design protection structures that stop most falls without damage, with the acceptance that an occasional large fall would cause damage requiring repair.

The probabilistic calculations require that the uncertainty in all the design parameters be quantified, expressed in terms of probability distributions. For each parameter, design values are selected based on knowledge of site conditions such as records of previous rock falls if available, and experience with similar rock fall locations. The simplest probability distribution is triangular defined by likely maximum, average, and minimum values; this is useful where the site information is insufficient to define a mean and standard deviation. For geological parameters where maximum and minimum values can be reasonably well defined, it may be preferable to avoid distributions that extend to infinity, as is the case with normal distributions, in order to avoid calculated high-impact energy events that have a very low probability of occurrence.

It is noted that probabilistic design methods can be used even if little information on design parameters is available; judgment can be applied to develop, for example, simple triangular distributions defining the likely range of values that may occur. This can be a useful exercise that quantifies the uncertainty in the site information.

Calculation for the probability distribution of the impact energy described in this section has been carried out using the program **@Risk 5.0** that is an add-on to Excel spreadsheets (Palisade Corp., 2012). The calculation procedure is to first define each input parameter, such as the rock fall mass and velocity, as a probability distribution. Second, the impact energy is calculated using Monte Carlo simulation to generate a probability distribution for the energy. Monte Carlo simulation involves running the analysis many times, with a random number being used to select, for each analysis, a value for each input parameter from their respective probability distributions. The particular usefulness of Monte Carlo analysis is that it is possible to combine input parameters with different types of probability distributions into a new distribution of the output parameter.

The application of the program **@Risk** to the calculation of a probability distribution for impact energy is illustrated, for a hypothetical site, as follows.

Topography—The rock face above the protection structure is 110 m (360 ft) high at an overall face angle of 60° to 65° degrees. While the overall rock face is at a uniform angle, in detail the face is highly irregular, comprising a series of ridges and gullies related to the geology as described below.

Geology—The rock exposed in the face is a very strong, tabular bedded limestone containing seams of weak shale. The spacing of the bedding planes is about 1 to 2 m (3.3 to 6.6 ft), and the shale seams are about 100 to 200 mm (4 to 8 in.) wide. The beds are oriented at right angles to the face, and dip vertically. The relatively rapid weathering of the shale has resulted in the formation of prominent limestone ridges on the face and gullies in the shale. The significant falls are in the relatively strong limestone.

The following is a discussion of values for the parameters used to find the probability distributions for the mass and velocity of the rock falls.

8.3.1 Probability Distribution of Rock Fall Mass

The rock falls are assumed to be discoid-shaped formed by slabs of limestone that have had their irregular edges broken off at impacts points during the fall. The dimensions are estimated from the bedding and joint spacing observed on the face, as well previous rock falls observed on the riverbank.

The dimensions of previous, commonly occurring falls have thicknesses of about 0.2 m (0.7 ft) and diameters of about 0.5 m (1.6 ft), for a volume of about 0.04 cu. m (0.5 cu. yd) and a mass of 100 kg (220 lb). The estimated dimensions of the largest falls that may occur are a thickness of 1 m (3.3 ft) and a diameter of about 2 m (6.6 ft), for a volume 3 cu. m (4 cu. yd) and a mass of about 8,000 kg (17,600 lb). These dimensions refer to individual rock falls.

The distribution for the mass of the rock falls is based on the knowledge that bedding and joints are commonly closely spaced and have low persistence, and that wide spacing and high persistence discontinuities are less common. Consequently, for blocks of rock formed by these discontinuities, small blocks are much more common than large blocks. These characteristics can be modeled as lognormal probability distributions as illustrated in Figure 8.3 (Wyllie and Mah, 2002).

Figure 8.3 shows the assumed lognormal distributions for the rock fall diameter and thickness for the discoid-shaped blocks. These distributions are defined by the average

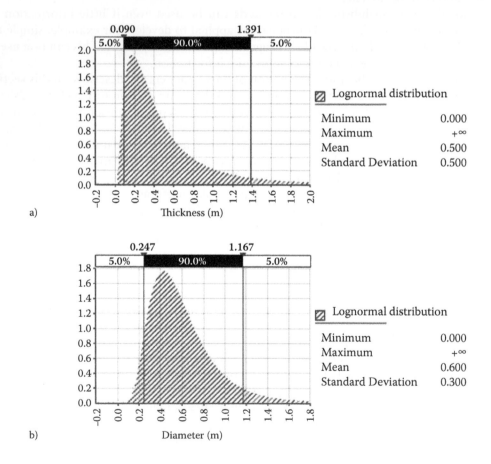

Figure 8.3 Lognormal distributions of rock fall dimensions, discoid shape. (a) rock fall thickness; (b) rock fall diameter.

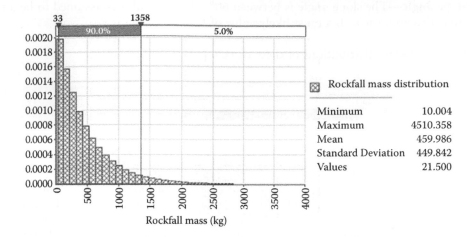

Figure 8.4 Distribution of rock fall masses for discoid-shaped blocks—output from Monte Carlo simulation.

thickness is 0.5 m (1.6 ft) with a standard deviation of 0.5 m (1.6 ft), and the average diameter is 0.6 m (2 ft) with a standard deviation of 0.3 m (1 ft). The mass of the rock fall is calculated from the volume and the unit mass of 2,650 kg · m^{-3} (165 lb · ft^{-3}). Figure 8.4 shows the calculated distribution of the rock fall mass, with 90% of the falls having masses of less than 1,360 kg (3,000 lb), and the average mass being 460 kg (1,000 lb). The largest calculated mass is about 4,500 kg (9,900 lb); this has a very low probability of occurring and is generated by the long tails of the lognormal distributions.

8.3.2 Probability Distribution of Rock Fall Velocity

The velocity v, of the rock fall is related to the fall height H, the slope angle, ψ_s, and the effective friction coefficient of the slope surface μ' according to the following relationship (see Section 3.2.2):

$$V = \sqrt{2 \cdot g \cdot H \left(1 - \frac{\mu'}{\tan \psi_s}\right)} \qquad (3.13)$$

The probability distribution of the velocity was developed by assuming the following values for these three rock fall parameters:

- **Fall height**—The maximum height is 110 m (360 ft), but rock falls can originate from any location on the face so the distribution of fall heights is uniform from 8 to 110 m (26 to 360 ft). The minimum height is 8 m (26 ft) because this is the height of the proposed protection structure.
- **Friction**—The effective friction coefficient depends on the strength of the slope material and the irregularity of the surface and has values between about 0.1 and 0.3 based on the information provided in Table 3.1. The friction coefficient distribution is modeled as a beta distribution with a range from 0.1 to 0.3 and a most likely value of about 0.16. The value of the beta distribution is that the end points (and realistic limits) of the distribution can be defined, and the mean can be skewed to reflect a nonuniform distribution of values (Harr, 1977).

- **Slope angle**—The slope angle is between 60° and 65° and was assumed to be a triangular distribution with a most likely value of 62.5°.

Figure 8.5 shows the distributions of these three input parameters.

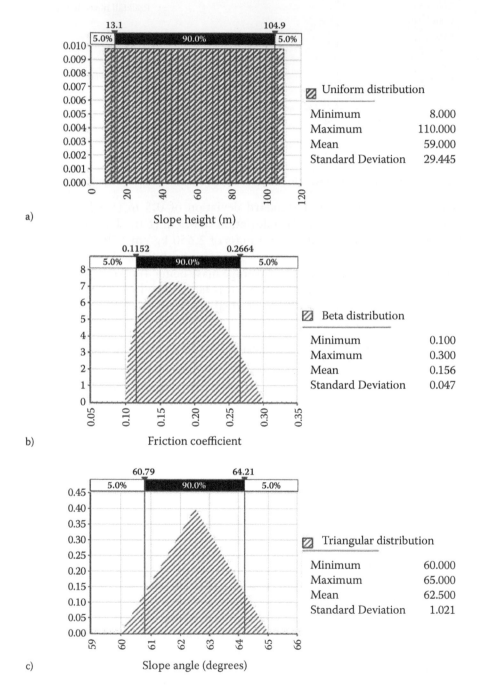

a) Slope height (m)

b) Friction coefficient

c) Slope angle (degrees)

Figure 8.5 Probability distributions for three input parameters used to calculate rock fall velocity. (a) Fall height—uniform; (b) friction coefficient—beta; (c) slope angle—triangular.

Figure 8.6 shows the calculated distribution of the velocity at the base of the slope using Equation (3.13); the velocities range from 10 to 42 m · s^{-1} (35 to 140 ft · s^{-1}), with an average value of 26.2 m · s^{-1} (85 ft · s^{-1})

The estimated probability distribution of the impact kinetic energy, ($E = 0.5 \, m \cdot V^2$) at the level of the protection structure is calculated using Monte Carlo simulation to combine the distributions for the mass and velocity shown in Figures 8.4 and 8.6, respectively.

Figure 8.7 shows the probability distribution for the impact energy in which most rock falls have energies less than about 200 kJ (75 ft tonf), and that 90% of the rock falls have energies less than 580 kJ (215 ft tonf). If the 90 percentile of the energy is selected as the service state energy, then three times this level, or 1,740 kJ (640 ft tonf), is the ultimate limit state energy (see Section 8.2). From the energy distribution plot it is found that an energy of 1740 kJ (640 ft tonf) occurs at about the 94th percentile.

In summary, the Monte Carlo analysis provides a method of quantifying the uncertainty of five input parameters, defined by four probability distributions, to obtain the probability distribution of the impact energy. The impact energy distribution allows a design energy to

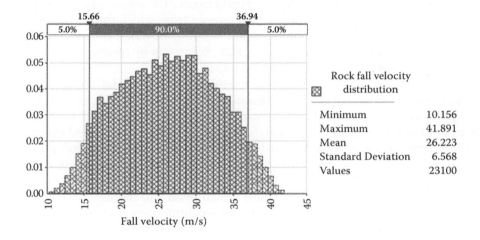

Figure 8.6 Probability distribution of rock fall velocities—output from Monte Carlo simulation.

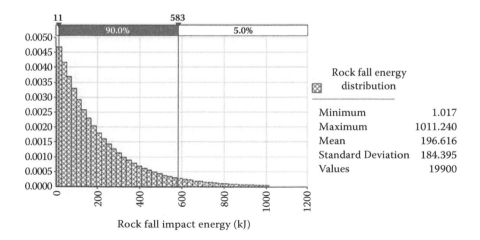

Figure 8.7 Probability distribution of impact kinetic energy—output from Monte Carlo simulation.

be selected that meets a required level of reliability, with the understanding that occasional large falls may damage the protection structure.

8.4 DETERMINATION OF ROCK FALL RETURN PERIODS

The design of rock fall protection structures may require values for the return periods, or frequencies, of rock falls of certain sizes for consistency with other project requirements such as seismic design. This section describes procedures to find the relationship between the masses of rock falls and their frequency of occurrence. Also discussed is the extrapolation of available information on rock fall frequency to obtain values for ultimate rock fall design mass as discussed in Section 8.2.

At rock fall study locations, it is common that histories of past rock falls can be ascertained from study of the site. That is, previous rock falls will usually be visible on the slope, and in the run-out area if it is undisturbed (see Figure 1.1). If the site is associated with a transportation system, an inventory of rock fall events may be available as part of a risk management program (see Section 8.5.2). Furthermore, the source area of rock falls can be inspected to measure joint spacing and persistence from which the likely range of block volumes can be determined. Other information that may be obtained from site inspections includes an understanding of rock fall paths and trajectories by observing impacts on trees and on the slope, and measuring run-out distances.

While site information will always be the starting point for rock fall studies, the design of protection structures may require an estimation of the largest rock fall that could occur over the design life of the structure, the size of which is greater than any that have been recorded in the past. This may involve a statistical extrapolation of field data to find the relationship between return periods of rock falls and their dimensions. This relationship can then be used to select an acceptable design frequency, such as a 2 cu. m (2.6 cu. yd) rock fall may have a return period of 20 years, but a 5 cu. m (6.6 cu. yd) rock fall may have a return period of 100 years. The selection of an acceptable design mass and frequency would depend on such factors as type of structure being protected—continuously occupied houses or a low-traffic road—and consistency with applicable design standards for the project return period.

This section describes two statistical methods for examining and extrapolating the relationship between the size of rock falls and their frequency of occurrence—the Gutenberg–Richter cumulative frequency method and the Gumbel extreme value theorem.

8.4.1 Gutenberg–Richter Cumulative Annual Frequency

The relationship between the volumes or masses of rock falls and their frequency of occurrence can be obtained by plotting the cumulative annual number of rock falls against the size of the falls on a log-log plot (Gutenberg and Richter, 1954; Hungr et al., 1999). An example of this type of plot is shown in Figure 8.8.

The data shown in Figure 8.8 were collected for the management of rock fall hazards in a transportation corridor in an area with steep mountainous terrain subject to severe winter climatic conditions. The geology comprised mainly strong, bedded limestone in which the types of rock falls ranged from ravelling of closely jointed seams to substantial falls of massive rock. The records provided information on the date and dimensions of 166 rock falls that occurred over a period of 18 years, and ranged in size from about 30 kg (65 lb) at 0.01 cu. m (0.01 cu. yd) to 26,500 kg (58,400 lb) at 10 cu. m (15 cu. yd). A technique that can be easily used by field personnel to visualize the dimensions of falls is to compare them with a number of standard sizes, e.g., baseball, basketball, desk, refrigerator, or car.

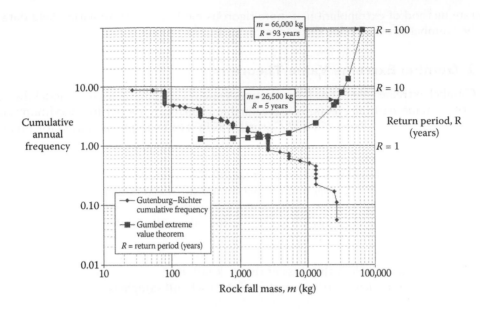

Figure 8.8 Plots of Gutenberg–Richter cumulative annual frequency and Gumbel extreme value theorem of rock fall mass against annual frequency/return period.

The cumulative annual frequency of rock fall masses for the 166 records in the database can be determined based on the annual frequency for any rock fall being 1/18 = 0.056. Table 8.1 shows the annual frequency of each fall and the cumulative frequency for the nine largest rock falls. Figure 8.8 shows the cumulative annual frequency plotted against the rock fall mass for all 166 recorded falls ranging in mass from 30 kg (65 lb) to 26,500 kg (58,400 lb).

The plot in Figure 8.8 shows a typical relationship between the size of a fall and its frequency of occurrence; that is, small falls occur more frequently than large falls. Also, the general gradient of the plot and the distribution of rock fall masses have been found to be similar to those for rock fall studies in other sites with similar topography and climate.

If an estimate were required of the cumulative annual frequency of larger falls than those observed in the field, it would be possible to extrapolate the plot by inspection. For example, it could be estimated by extrapolation that the annual frequency for a 30,000 kg (66,100 lb) rock fall would be about 0.03 years, or every 33 years. While, this is a rapid and reasonably

Table 8.1 Sample of rock fall data for the nine largest falls collected over 18 years, showing Gutenberg–Richter relationship between mass of the fall and cumulative annual frequency of falls

Rock fall mass (kg)	Annual frequency	Cumulative annual frequency
26,500	0.056	0.056
26,500	0.056	0.111
24,500	0.056	0.167
13,000	0.056	0.222
13,000	0.056	0.278
13,000	0.056	0.333
13,000	0.056	0.389
13,000	0.056	0.444
10,500	0.056	0.500

Note: Total number of falls = 166.

accurate method of extrapolation, a more rigorous method of extrapolating field data is to use the Gumbel extreme value theorem as described in Section 8.4.2.

8.4.2 Gumbel Extreme Value Theorem

The Gumbel extreme value Type I distribution, which is commonly used to model the occurrence of extreme natural events, is a reasonable probability model for maximum annual values of these events (Benjamin and Cornell, 1979). For example, it is used for prediction of maximum floods, maximum windstorm speeds, and other phenomena for which only the maximum design parameter is required.

Development of a Gumbel distribution requires selection, for each year of the site data, only a single data point representing the maximum for the year. The Gumbel distribution then finds, from a fit to the data, an estimate that represents a population of the maximum rock size for any year in the future. The steps in the calculation of the rock fall return period are as follows:

- The random variable is the mass of the rock fall, m.
- The probability density function $f(m)$ of the rock fall sample is

$$f(m) = \frac{1}{\beta} e^{-z - e^{-e}} \tag{8.1}$$

where

$$z = \left(\frac{m - \alpha}{\beta} \right),$$

α is the location parameter equal to: $\alpha = (M - 0.577\,\beta)$, and β is the scale parameter equal to:

$$\beta = S \frac{\sqrt{6}}{\pi}$$

M is the mean of the rock fall masses, S is the standard deviation of this data, and the term 0.577 is the Euler–Mascheroni constant. The function $f(m)$ has a range of minus infinity to plus infinity.

- The cumulative distribution function $F(m)$ of the probability density function is

$$F(m) = e^{-e^{-z}} \tag{8.2}$$

- The term $(1-F(m))$ represents the probability of exceedance of a rock fall with a given mass in one year.
- The reciprocal $[1/(1-F(m))]$ is the return period, R in years.

For the rock fall data of 166 events over 18 years discussed in Section 8.4.1 and plotted on Figure 8.8, a Gumbel extreme value set of calculations has been prepared so that the Gumbel distribution and Gutenberg–Richter cumulative frequency values can be compared. The Gumbel analysis involved first sorting the data by year, and then finding the maximum rock fall mass for each of the 18 years of records. These 18 records of the mass were then sorted in numerical order and the return period for each were calculated using Equations

(8.1) and (8.2) following the procedure shown on Table 8.2. In addition to the 18 recorded rock falls, the predicted return periods were calculated for three potential falls with volumes of 12 cu. m (15 cu. yd), 15 cu. m (20 cu. yd), and 25 cu. m (35 cu. yd) in order to extrapolate the recorded field data.

The parameters used in the analysis are as follows:

For the 18 largest falls for each year, plus blocks with volumes of 12, 15, and 25 cu. m the average mass, $M = 12{,}755$ kg and standard deviation, $S = 17{,}365$ kg. The applicable rock unit mass is 2650 kg $\cdot$ m^{-3}.

Location factor, $\alpha = 4{,}949$ and scale factor, $\beta = 13{,}540$.

The rock fall mass is plotted against the return period on Figure 8.8 in order to compare the Gutenberg–Richter and Gumbel distributions. For rock falls greater than the largest recorded fall of 26,500 kg (58,400 lb) for which the return period is about 5 years, the return period for a 66,000 kg (146,000 lb) rock fall is about 100 years.

It is possible to assess the reliability of the calculated Gumbel return periods using the coefficient of variation, COV of the data for the 166 event data set which has an average, M of 1,635 kg and a standard deviation, S of 4,305 kg.

$$\text{COV} = S/M = 2.63 \tag{8.3}$$

Table 8.2 Annual maximum rock fall volume and mass for 18 years of records showing results of Gumbel extreme value theorem to calculate return periods; calculated return periods also shown for volumes of 12, 15, and 25 cu. m (15, 20, and 35 cu. yd)

Volume (cu. m)	Mass, m (kg)	$z = \dfrac{(m-\alpha)}{\beta}$	Cumulative distribution function, F(z)	Annual probability of exceedence (1-F(z))	Return period, years R = 1/(1-F(z))
0.1	265	−0.346	0.243	0.757	1.32
0.1	265	−0.346	0.243	0.757	1.32
0.1	265	−0.346	0.243	0.757	1.32
0.1	265	−0.346	0.243	0.757	1.32
0.1	265	−0.346	0.243	0.757	1.32
0.3	795	−0.306	0.257	0.743	1.35
0.5	1,325	−0.267	0.271	0.729	1.37
0.7	1,855	−0.228	0.285	0.715	1.40
0.76	2,014	−0.216	0.289	0.711	1.40
1	2,650	−0.169	0.306	0.694	1.44
2	5,301	0.026	0.378	0.622	1.61
2	5,301	0.026	0.378	0.622	1.61
2	5,301	0.026	0.378	0.622	1.61
5	13,252	0.614	0.582	0.418	2.39
5	13,252	0.614	0.582	0.418	2.39
9.3	24,648	1.456	0.792	0.208	4.81
10	26,504	1.593	0.816	0.184	5.43
10	26,504	1.593	0.816	0.184	5.43
12	31,804	1.984	0.872	0.128	7.78
15	39,755	2.571	0.926	0.074	13.59
25	66,259	4.529	0.989	0.011	93.20

This value of the coefficient of variation is large and is the result of the wide range for the largest annual rock fall volumes—four orders of magnitude from 0.01 cu. m (0.01 cu. yd) to 10 cu. m (15 cu. yd). This range of volumes is usual for rock fall events where small falls are common, and large falls occur infrequently. The reliability of the data is also limited by the short period of data collection of 18 years. Because of the high value for COV for this data set and short period of data collection, the Gumbel calculation of return periods are approximate. In comparison, Gumbel calculations for floods, for example, may be more reliable because it is expected that the range of maximum annual flood events would be narrower and the flood levels could be more precisely measured than the volume of rock falls.

8.5 RISK MANAGEMENT OF ROCK FALL HAZARDS

The basic requirement for installing a rock fall protection structure is to provide a level of protection that is consistent with the probable consequences of an accident. For example, a suitable protection method for a continuously occupied house located below an active source of rock falls would be a substantial ditch or fence that could be inspected and maintained. In contrast, for the same slope, above a lightly used industrial road that is not open to the public, it may be decided that minor protection such as draped mesh, or even no protection, is justified because the users are aware of the hazard and can take evasive action, as required.

The procedures for selecting appropriate protection measures can be quantified using risk management and decision analysis as described in this section (Wyllie, 2006). In summary, this approach involves developing the following qualitative data:

- Exposure risk, or the probability that a rock fall will result in an accident.
- Consequences of an accident in financial terms such as damage to infrastructure or vehicles, as well indirect costs such as traffic delays.
- Costs of constructing and maintaining protection measures.

8.5.1 Definitions of Hazard and Risk

In quantifying the effects of natural hazards such as rock falls, it is necessary to define **hazard** and **risk**, the two separate, but related, components of these events, as described below.

Hazard—A rock fall hazard is a combination of a source, a triggering event, and a path from the source to the at-risk object (Figures 8.9 and 8.10).

The **source** may be an excavated rock face or natural slope and is defined by the topography and geology of the site. With respect to topography, the slope angle must be steeper than the angle of repose (about 37°) in order to generate rock falls (see Figure 1.1). Also, the higher the slope, the greater the surface area that can be the source of falls, and the greater the velocity, up to an approximate terminal velocity, that rock falls can attain. The geological factors that directly influence rock fall hazards are the characteristics of the discontinuities, with the spacing and persistence defining the size of rock falls, and their orientation defining the likelihood of blocks sliding or toppling from the face. The rock strength is also an important issue since blocks of strong rock tend to remain intact as they roll and bounce down the slope and not break down into smaller, less hazardous fragments.

A rock fall will result from a *triggering event*, often related to the climate or seismicity. As discussed in Section 1.3, records of rock fall events show that rock fall frequency is strongly correlated with rainfall and freezing temperatures because water and ice pressures acting in cracks loosen and displace blocks of rock (Figure 1.4). A related cause of falls in wet climates is the growth of tree roots in cracks in the rock (Figure 1.5). Another important triggering

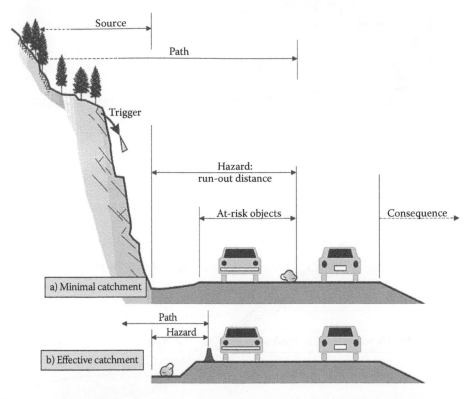

Figure 8.9 Definition of hazard and risk for a highway below a rock slope; rock fall protection comprises a ditch at the base of the slope, (a) minimal catchment allows rock to reach traveled surface, and (b) effective catchment contains falls.

mechanism in seismic areas is strong ground motions; many rock falls and landslides were triggered, for example, by the 1994 Northridge earthquake near Los Angeles (see Section 1.5) and the 2011 Christchurch earthquake in New Zealand (Dellow et al., 2011) (Figure 3.1).

The ground surface between the source and the at-risk object is the *path*, which may include the rock face, talus slopes, ditches, and developed areas such as roads and houses as shown in Figures 8.9 and 8.10. The distance that rock falls may travel along a pathway depends on such factors as the size of the block, the inclination of the slope, and the composition and irregularity of the surface. For example, the travel distance is likely to be less on talus and soft soil than for asphalt and bare rock. Protection measures may be located at any position along the path, but preferably in a location that facilitates construction and maintenance. Impact energies may also diminish with distance traveled since blocks tend to break into smaller fragments with successive impacts.

Risk—any object that is on the path of a potential rock fall, is at risk for damage. For example, in Figure 8.9(a) where the ditch is shallow with a rounded base, rock falls may roll through the ditch and on to the road where a risk exists that they could damage vehicles; the risk to vehicles increases with increasing traffic frequency.

Another component of risk is the consequence of an accident. For example, if the slope below the road is steep with no traffic barrier, the consequence of an accident will be more severe than if the ground were level. The consequence of an accident would be one of the factors taken into account when evaluating the need for protection measures.

In Figure 8.10(a), a house located at the base of the talus slope is at risk from damage by rocks that roll beyond the talus into the "rock fall shadow." The "shadow" is defined

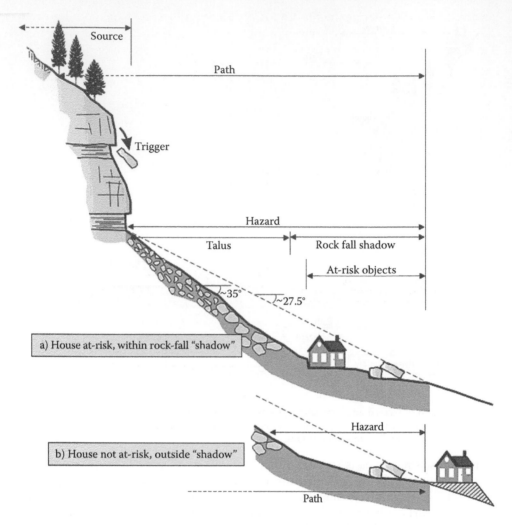

Figure 8.10 Hazard and risk zones for a house below a rock slope and talus deposit. (a) house at risk within rock fall shadow, and (b) house at safe location outside rock fall shadow.

by a line at an angle of about 27.5° below the horizontal from the crest of the talus slope to intersect the ground surface beyond the talus. The risk would be increased if the house were on the talus where more rock falls accumulate compared to the shadow zone (see also Figure 1.1).

Figures 8.9(b) and 8.10(b) show how the risk can be diminished or eliminated when the at-risk object is outside the rock fall path. In Figure 8.9(b), excavation of the ditch and installation of a barrier to contain falls limits the path to the outer edge of the ditch and reduces the risk of falls reaching the road. In Figure 8.10(b), the house is beyond the "shadow" zone and outside the limit of the rock fall path.

8.5.2 Inventories of Hazard and Risk

The first step in quantifying hazard and risk is to make an inventory of site conditions. For example, on transportation routes with a large number of rock slopes, the inventory would describe the physical characteristics of each slope to define the hazard, as well as the path and traffic conditions that define the risk. A well-known inventory method, known as the Rock Fall Hazard-Rating System, which is applicable to highways (Wyllie, 1987; Pierson et al., 1990), assigns scores ranging from 3 points to 81 points to each of nine site parameters (see Table 8.3, as modified). The site parameters are categorized as either *hazard factors* related to slope conditions, or *risk factors* related to path and traffic conditions, as follows:

- *Hazard factors*
  - Slope height—maximum height of rock fall sources
  - Geology—stability controlled by either structural geology or differential weathering
  - Block size—controlled by joint spacing and rock strength
  - Climate—effects of both precipitation and freezing temperatures
  - Rock fall history—record of past events
- *Risk factors*
  - Average vehicle risk—percent of time that vehicles are below the slope
  - Sight distance—ability of drivers to see rock fall on road and take evasive action
  - Roadway width—space available to contain falls and for drivers to take evasive action
  - Ditch effectiveness—width and depth of ditch related to slope height and angle defines efficacy of ditch to contain falls

Each site parameter is assigned a score according to the arithmetic range of 3, 9, 27, or 81, with 3 for favorable conditions such as a wide ditch, and 81 for severe conditions such as a record of very frequent rock falls.

Since the hazard and risk factors are independent, separate numeric scores for the hazards and risks can be obtained by adding the scores for each parameter. The final score for the site is the product of the hazard and risk scores. This information can be used to rank a large number of rock slopes to identify the most hazardous locations, and to prioritize remedial programs, with sites having high scores being remediated before those with lower scores.

The following example illustrates how the product of the hazard and risk scores identifies a hazardous location and how remedial work can lower the score.

Hazard rating for a rock slope with the following characteristics using Table 8.3:

1. Slope height—height of 25 m (82 ft), score = 27
2. Structural geology—rock contains persistent, planar joints with no infilling that dip out of the face on which planar-type instability can occur, score = 27
3. Block size—joint spacing is about 0.5 to 1 m (1.6 to 3.3 ft) forming blocks with volumes up to about 0.5 cu. m (0.7 cu. yd), score = 9
4. Climate—wet climate with occasional periods of freezing temperatures during the winter, score = 27
5. History—very frequent rock falls, score = 81
     Rating = (27 + 27 + 9 + 27 + 81) = 171;

Risk rating for a road with the following characteristics using Table 8.3:

1. Vehicle risk, low-volume road with vehicle occupancy under slope about 20% of time, score = 3
2. Sight distance, score = low curvature with adequate sight distance to avoid rock falls, score = 3
3. Road width—width of 8.5 m (28 ft) with narrow shoulders, score = 27

Table 8.3 Summary sheet of rock fall hazard-rating system

Category	Hazard-rating criteria and scores			
	Points 3	Points 9	Points 27	Points 81
1. Slope height (m)	7.5 m (25 ft)	15 m (50 ft)	23 m (75 ft)	30 m (100 ft)
2. Geologic character				
Case 1				
Structural condition	Discontinuous joints, favorable orientation	Discontinuous joints, random orientation	Continuous joints, adverse orientation	Continuous joints, adverse orientation
Rock friction	Rough, irregular	Undulating	Planar	Clay infilling, or slickensided
Case 2				
Structural condition	Few differential erosion features	Occasional erosion features	Many erosion features	Major erosion features
Difference in erosion rates	Small difference	Moderate difference	Large difference	Extreme difference
3. Block size	0.3 m (1 ft)	0.6 m (2 ft)	1.0 m (3.3 ft)	1.2 m (4 ft)
Quantity of rock fall event	3 m³ (4 yd³)	6 m³ (8 yd³)	9 m³ (12 yd³)	12 m³ (16 yd³)
4. Climate and presence of water on slope	Low to moderate precipitation; no freezing periods, no water on slope	Moderate precipitation, or short freezing periods, or intermittent water on slope	High precipitation or long freezing periods, or continual water on slope	High precipitation and long freezing periods, or continual water on slope and long freezing periods
5. Rock fall history	Few falls	Occasional falls	Many falls	Constant falls

Category	Risk-factor rating criteria and scores			
	Points 3	Points 9	Points 27	Points 81
1. Ditch effectiveness	Good catchment	Moderate catchment	Limited catchment	No catchment
2. Average vehicle risk (% of time)	25% of the time	50% of the time	75% of the time	100% of the time
3. Percentage of decision sight distance (% of design value)	Adequate sight distance, 100% of design value	Moderate sight distance, 80% of the design value	Limited sight distance, 60% of design value	Very limited sight distance, 40% of design value
4. Roadway width including paved shoulders (m)	13.5 m (44 ft)	11 m (36 ft)	8.5 m (28 ft)	6 m (20 ft)

4. Ditch effectiveness—narrow ditch that only contains falls from the lower one-third of the rock face, score = 27

> Rating = (3 + 3 + 27 + 27) = 60
> **Total rating** = (171 × 60) = 10,260.

The actual value of the total rating is less important with respect to rock fall hazards than the relative totals of all the sites included in the inventory.

An application of the hazard scoring system is as follows. If the hazard at the site in the example described above is reduced by widening the ditch so that it will be highly effective in containing rock falls, then the ditch score is reduced from 27 to 3, and the risk rating is reduced from 60 to 36. While the ditch improvements reduce the rock fall risk at the site, this work does not change the hazard factors because rock falls will still occur.

The total rating for the site after excavating the ditch is (171 × 36) = **6,156**, a reduction of 40%, and a possible change in the hazard ranking of the site. The numerical hazard rating of rock slopes is an effective means of making a qualitative inventory of stability conditions, and of keeping records of remedial work.

The alternative method of obtaining the hazard rating is to add the 10 hazard and risk ratings, which for the example discussed above would be (171 + 60) = 231 before excavation of the ditch. After excavation of the ditch the total score would be (171 + 36) = 207, which is a reduction in the score of only 10% that does not clearly identify the significant reduction in the rock fall hazard that has been achieved.

8.5.3 Probabilities of Rock Falls

The results of the slope inventory discussed in Section 8.5.2, together with available rock fall records, can be used to calculate the probability of rock fall occurrence, as demonstrated below.

For a transportation system located in steep mountainous terrain where the winters are cold with many freeze-thaw cycles, and periods of intense rainfall occur, the following information was obtained from the rock fall inventory.

- Number of rock falls   = 682
- Number of years of data   = 31
- Number of rock cuts   = 118

Probabilities of rock falls can be calculated from these figures as follows.

- Annual probability of rock fall on each rock cut, $p_r = (682/(31 \times 118)) = 0.19$ falls/ rock cut/year;

This probability value represents the average rock fall hazard over this alignment length, and is independent of traffic conditions. A p_r value of 0.19 falls per cut per year means that, on average, a rock fall occurs every 5 years on each cut. This rock fall probability would be a baseline value that could be compared, for example, with the frequency of events in the future to assess whether stability conditions were deteriorating, as well as determining the efficacy of stabilization work. Furthermore, the rock fall frequency in this area could be compared to the frequency in another area to assess the relative hazard.

8.5.4 Calculation of Relative Risk

Having calculated the rock fall hazard for the rock slopes as discussed in Section 8.5.3 above, the next step is to calculate the risk to facilities below the slope that are located on the rock fall path.

Figure 8.11 represents a 250 m (820 ft) long rock face, in profile, with four possible at-risk objects within the rock fall path at the base of the slope. The four objects are a house (stationary) that occupies a slope length of 20 m (65 ft), a logging truck traveling at 20 km per hour with a traffic count of 15 per day, a freight train traveling at 40 km per hour with a traffic count of 50 per day and a vehicle traveling at 100 km per hour with a traffic count of 6,000 per day.

The risk of driver impact, r_e is the proportion of time per day that drivers are exposed to rock falls, which can be calculated as follows:

$$r_e = \frac{L \cdot N}{S(24,000)} \tag{8.4}$$

where L is the length of the cut (m), N is the traffic count (vehicles/day), and S is the vehicle speed in km $\cdot$ hr^{-1}. Equation (8.4) shows that the driver exposure time increases with the length of the cut and the traffic count but decreases with greater speed.

The application of equation (8.4) to the four objects at risk gives the following values for the relative exposure risk:

House	r_e	= 1
Logging truck	$r_e = (250 \cdot 15)/(20 \cdot 24\ 000)$	= 0.008
Freight train	$r_e = (250 \cdot 50)/(40 \cdot 24\ 000)$	= 0.013
Car	$r_e = (250 \cdot 6\ 000)/(100 \cdot 24\ 000)$	= 0.625

These values of r_e show that for the moving objects the cars having the highest relative risk because of the high traffic count, compared with trains that operate at lower speeds (and longer exposure times) but have a lower traffic count. The house, located within the rock fall shadow, is continuously at risk ($r_e = 1$).

The calculation just examines the risk to the driver of the vehicle (length = 1 m), and is not concerned with the vehicle being impacted.

The relative probability of each these objects being impacted by a fall can be calculated for the annual probability of a rock fall on this slope being 0.19. The product of the exposure risk r_e and the rock fall probability p_r, is the annual probability of an object being impacted by a rock fall, p_i given by:

Probability of impact, $p_i = r_e \cdot p_r$ (8.5)

Figure 8.11 lists the calculated p_i values, using Equation (8.5) for each vehicle that are equivalent to the relative impact risk for each type of object. The probability value for the house takes into account that the 20 m (65 ft) length of the house extends over only 8% of the 250 m (820 ft) long cut (i.e., $0.19 \cdot 20/250 = 1.5\%$). It is also assumed that the house is occupied year round; the risk would be diminished if it were only occupied part of the year.

The calculations of risk and probability of impact shown on Figure 8.11 represent the maximum risk from rock falls to which objects are subjected. In reality, the actual risk is reduced by a number of site specific factors that may include:

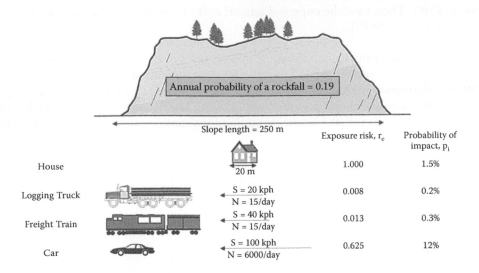

Figure 8.11 Calculation of relative risk of four objects below the slope that are on the rock fall path.

- Rock falls from the lower part of the slope will probably only travel a limited distance along the pathway and not reach the traffic lanes. Falls can be expected to originate from any height on the slope with equal probability (see Figure 8.5a));
- Small rock falls that cause little or no damage are more frequent than large falls (see Figure 8.4). Furthermore, trains are not damaged by small rock falls that could severely damage a car;
- For rock falls that are on the road in front of the vehicle, avoidance actions such as swerving or braking can be taken. However, in the case of trains, their inability to swerve and stopping distance of up to 1.5 km (0.9 mile), means that they are likely to impact any rock fall on the track.

8.5.5 Decision Analysis—Selection of Optimum Mitigation

Decision analysis, which is a well-proven technique for making decisions under uncertainty, provides a method of evaluating mitigation options (Raiffa, 1968; Wyllie et al., 1979). The basic principle of decision analysis is to determine, for each strategy, the expected cost, EC, that is defined as:

$$EC = \left(C_c + \sum_1^i (p_i \cdot C_i) \right) \tag{8.6}$$

where C_c is the construction cost of remedial work, and for each of type of incident (rock fall) i, p_i is the annual probability of a rock fall occurring and C_i the cost of that incident. That is, the expected annual cost of a rock fall is the product of the probability of occurrence of the event (p_i) and the expected cost of the event (C_i); expected annual cost, $EC = (p_i \cdot C_i)$.

Figure 8.12 shows a "decision tree" where the probability of an event occurring, and its cost, are shown for three mitigation strategies, S_A, S_B, and S_C. The calculated expected cost, EC, for each strategy is shown on the diagram. For example, for existing conditions (strategy A), the annual probability of a minor rock fall is 0.6, and the cost of this event, if it occurs, is $200. Therefore, the expected annual cost of a minor fall for existing conditions is ($EC = 0.6 \cdot 200 = \$120$).

For all three strategies, the same type of chance incidents can occur as indicated by the circle (O). These incidents comprise first, a minor rock fall that causes no damage or delays to traffic and can be removed by regular maintenance operations; second, a delay to traffic as well as stabilization of the rock fall area; and third, an impact that results in damage and injury, stabilization work, and possibly legal action. The average cost C_i, of each of these three types of incidents is listed below, assuming for simplicity that these costs are constant for each strategy.

Minor falls $C_{A,1}$ = $200
Delay, stabilization $C_{A,2}$ = $10,000
Impact, delay, stabilization $C_{A,3}$ = $1,000,000

The rock fall records show the probability of occurrence of each of type of incident for existing conditions strategy A is

Minor falls $p_{A,1}$ = 0.6
Delay, stabilization $p_{A,2}$ = 0.3
Impact, delay, stabilization $p_{A,3}$ = 0.1

Therefore, the total expected annual cost of all events for existing conditions is (EC_A = 120 + 3000 + 100,000 = $103,120). The total expected cost for each strategy is the total cost of the uncertain events that may occur plus the cost of preventing, or reducing the probability, of these events.

The decision tree in Figure 8.12 shows the structure of a decision involving the following three strategies, with a square (□) indicating the decision point:

Strategy S_A—existing conditions with no mitigation
Strategy S_B—stabilization with rock bolts and shotcrete
Strategy S_C—construction of rock fall shed

For each strategy, the corresponding construction cost (C_{cn}) is

- C_{cA} = 0—no mitigation
- C_{cB} = $50,000—rock bolts and shotcrete
- C_{cC} = $1.5 million—rock fall shed

The probability of occurrence of each incident for each strategy is determined by a combination of rock fall records and judgment of the likely performance of the mitigation measure. For example, on Figure 8.12, the probability of a minor fall for existing conditions is 0.6 (60%), and it is estimated that the slope stabilization work will change the probability of minor falls to 0.9 (90%) of all events, and construction of the shed will change the probability of minor falls to 0.97 (97%) of all events. The corresponding probabilities of major events are 10%, 1%, and 0.5% for the three mitigation strategies.

An important property of the probability values is that the total probability at each chance point is 1.0 because this is the sum of all the incidents that can occur. For existing conditions, information on frequency of impacts and delays may be available, as well as rock falls that result in no disruptions to traffic. The p_i values for the two mitigation strategies can be controlled to some extent by their design. For example, spot bolting would only reduce

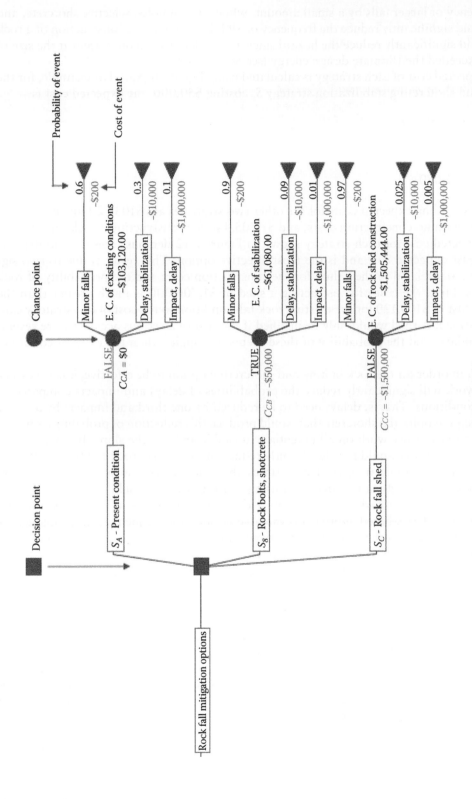

Figure 8.12 Decision tree showing three rock fall mitigation strategies and three possible types of rock fall events with the probability of occurrence and cost of each event. "True" is most favourable decision.

the frequency of larger falls by a small amount, while pattern bolts, selective shotcrete, and mesh would significantly reduce the frequency of all falls. In contrast, construction of a rock shed would significantly reduce the hazard since an accident would only occur if the size of the fall exceeded the Ultimate design energy (see Section 8.2).

The expected cost of each strategy is calculated using Equation (8.6). For example, for the bolting and shotcreting stabilization strategy S_B costing \$50,000, the expected cost is

$$EC = C_{cB} + \Sigma(p_{i,n}C_{i,n})$$

$$= \$50,000 + (0.9 \cdot \$200 + 0.09 \cdot \$10,000 + 0.01 \cdot \$1,000,000)$$

$$= \$61,080$$

The corresponding expected costs for the other two strategies are \$103,120 for the existing conditions with no construction costs, and \$1,505,444 for constructing a rock fall shed.

The expected cost for each strategy shown in Figure 8.12 demonstrates that Strategy S_B has the least expected cost and is the most effective option. This result is due to two significant factors related to the values for the construction costs and the probability of rock fall events. First, the cost of constructing a shed of \$1,500,000 is much greater than the expected cost of \$103,120 for the existing rock fall condition so the cost of this strategy cannot be justified. A shed would only be justified if the cost of a serious accident were several million dollars, and the probability of these events was higher than the existing condition of 0.1.

Second, in order for the rock bolting and shotcreting option to be effective, it is necessary that the work will significantly reduce the probabilities of delays and impacts compared to existing conditions. That is, delays need to be reduced by one third and impacts by a factor of 10. Design of bolts and shotcrete that would produce this reduction of probabilities would likely require extensive work on all potentially unstable areas of the slope. In comparison, localized spot bolting would not significantly reduce the long-term rock fall frequency.

In summary, decision analysis demonstrates that for facilities such as transportation routes, rock fall hazards can be managed using a risk-based approach. This optimizes the costs of stabilization or rock fall protection with respect to the cost of accidents resulting from rock falls. A risk-based approach is valuable because the frequency and consequences of rock falls, as well as the costs and effectiveness of mitigation, are uncertain, but these uncertainties are quantified in decision analysis.

Some of the benefits of applying risk management programs in implementing rock fall mitigation programs are

- The implementation of a consistent, ongoing mitigation program may provide, in North America, some legal protection against negligence for owners of facilities from lawsuits brought by victims of rock fall accidents.
- Mitigation programs can be optimized by preparing an inventory of rock slopes together with the hazard and risk at each site. This information, together with rock fall probability values and decision analysis, can be used to select and design the most economically effective program.
- For facilities with many rock slopes and climatic conditions where rapid rock weathering occurs, it may be necessary to undertake long-term mitigation programs in which remediation work on some slopes is required every 5 to 10 years. Risk management will be useful in developing a long-term strategy for the program. For example, for slopes where rapid deterioration of stability conditions is occurring, decision analysis

would help to identify a cost-effective mitigation measure for long-term stability. That is, installing pattern bolts and shotcrete may be more economical than excavating a ditch that needs to be cleaned every few weeks or months.

Chapter 9

Design Principles of Rock Fall Protection Structures

The design of rock fall protection structures such as wire-rope fences is based on the efficient absorption of impact energy. This requires that the structure be both flexible and stiff. That is, the structure must be flexible to deflect during impact, and also stiff so that energy is absorbed during deflection. The most effective protection structures are those in which energy is absorbed uniformly throughout the impact period. This chapter discusses the design principles for protection structures that absorb energy efficiently.

9.1 STRUCTURE LOCATION WITH RESPECT TO IMPACT POINTS

One of the factors in the location of protection structures is to find a position where the rock fall energy is relatively low. That is, in order to limit the impact energy on the structure, it should be located just after an impact point to benefit from the loss of energy due to plastic deformation that occurs during the time that the rock is in contact with the ground.

Energy loss during impact can be demonstrated for rock falls at the Ehime test site in Japan. For one of the tests, Figure 2.6 shows the impact and restitution velocities at each impact, and the energy partition plot in Figure 6.5 shows the corresponding kinetic and rotational energies over the full extent of the rock fall path. The energy partition plot shows how the kinetic energy increases due to gravitational acceleration during the trajectories and is lost during the impacts, while the rotational energy changes (either increases or decreases) during the impacts, but remains constant during the trajectories. For impact #5 at a fall height of 15.5 m (50 ft), the impact kinetic energy is 53.4 kJ (20 ft tonf) while the restitution kinetic energy is 24.5 kJ (9 ft tonf), a loss of 28.9 kJ (11 ft tonf) or 54% during impact.

While this energy loss demonstrates the value in locating the structure immediately after impact points, the generally uniform slope topography at the Ehime test site means that impacts will not occur at well-defined locations on the slope. Therefore, no optimum location for a barrier on the slope can be found.

A common topographic feature that defines an impact point along a rock fall path is a break in slope forming a relatively shallow angle bench on which rock falls will tend to accumulate. If such a feature does not occur naturally, it may be worthwhile to excavate a bench on which to locate the barrier since it could be designed for a lower-impact energy than one located on the uniform slope.

For the rock fall site at Tornado Mountain discussed in Section 2.2.2 and shown in Figure 2.8, an 8 m (25 ft) wide bench was excavated in the slope on which the railway was constructed; Figure 9.1 shows a detail of the bench. Both the documented rock falls at the site impacted this bench and stopped within 30 m (100 ft) of the railway after falling over a slope distance of about 700 m (2,300 ft) down the uniform slope above the railway.

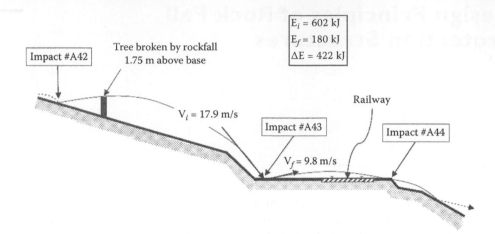

Figure 9.1 Tornado Mountain rock fall site—for impact #A43 on 8 m (25 ft) wide bench excavated for the railway, 70% of the impact energy is lost during impact.

Trajectory calculations for Boulder A show that for rock falls impacting the horizontal bench (impact #A43), the impact velocity was 17.9 m · s⁻¹ (60 ft · s⁻¹) and the restitution velocity was 9.8 m · s⁻¹ (30 ft · s⁻¹), representing a loss of kinetic energy of 422 kN (50 tonf) for the 3,750 kg (8,270 lb) block (i.e., $KE_{A43i} = 0.5 \cdot 3{,}750 \cdot 17.92^2 = 602$ kJ and $KE_{A43f} = 0.5 \cdot 3{,}750 \cdot 9.8^2 = 180$ kJ). That is, 70% of the impact energy was lost during impact #A43, whereas for impacts on the more uniform slope higher on the mountain, the typical energy loss was about 20%. This demonstrates the effectiveness of a horizontal bench on mitigating rock fall hazards; for a fence or barrier located along the outer edge of such a bench, the design energy can be 50% of the energy required on the uniform slope.

9.2 ATTENUATION OF ROCK FALL ENERGY IN PROTECTION STRUCTURES

Figure 9.2 illustrates two examples of rock fall barriers: a flexible wire-rope fence that has been effective in stopping and containing rock falls, and a rigid concrete wall that has been shattered by a rock fall. This section demonstrates how protection structures can be designed to attenuate and dissipate a portion of the impact energy rather absorb the entire energy, and how impact mechanics can be used to develop these designs.

9.2.1 Velocity Changes during Impact with a Fence

If a rock fall is stopped by a protection structure, then all the impact energy is absorbed in the structure because the impact translational and rotational velocities are reduced to zero. However, if the rock is redirected by the structure, then the restitution velocities have finite values and the difference between the impact and restitution velocities represents the portion of the impact energy absorbed in the structure (Figure 9.3).

Figure 9.3 shows a rock fall fence on a slope at two orientations—normal to the slope (Figure 9.3(a)), and inclined upslope (Figure 9.3(b)). The fence is impacted by a rock fall with initial translational velocity v_i and rotational velocity ω_i, at an impact angle of the rock with the net of θ_i.

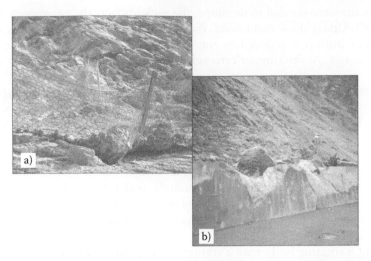

Figure 9.2 Behavior of flexible and rigid structures. (a) Flexible steel cable net that stops rock falls by deflection with no plastic deformation of the steel; (b) rigid concrete wall shattered by rock fall impact.

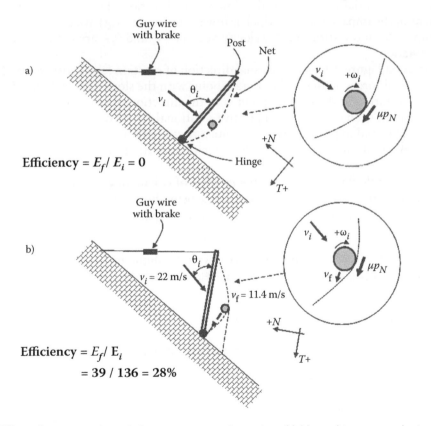

Figure 9.3 Effect of impact angle with fence on energy absorption. (a) Normal impact results in the fence absorbing all impact energy; (b) oblique impact results in rock being redirected off the net with partial absorption of impact energy. (See Worked Example 9B.)

For the fence oriented normal to the slope, the rock impacts the net approximately at right angles ($\theta_i = 90°$). Under these conditions, the rock deflects the net and its velocity progressively diminishes until it is reduced to zero at the point of maximum deflection before it rebounds off the net. At the time of zero velocity, all the impact energy has been absorbed by the net.

In contrast to the normal impact shown in (Figure 9.3[a], Figure 9.3[b]) shows the fence oriented upslope so that the rock fall impacts the net at an oblique angle ($\theta_i < 90°$). For this impact angle, the rock is redirected by the net and the velocity decreases to v_f when it loses contact with the net; at no time during the impact does the velocity become zero. For these conditions, only a portion of the impact energy is absorbed by the fence.

The behavior of a rotating body impacting a net at either a normal or oblique angle as shown in Figure 9.3 can be modeled approximately, using the principles of impact mechanics discussed in Chapter 4. That is, Equations (4.20) to (4.24) can be used to calculate the influence of the fence configuration on changes in the translational and angular velocities and angles during impact.

For impact of a rotating body with a stationary net, the parameters governing the impact behavior are the normal coefficient of restitution e_N, the radius of the body r, and its radius of gyration, k. In addition, the calculations require that normal and tangential axes be set up, with the positive directions of each defined, as well as the impact and restitution angles of the velocity vectors relative to the fence. Figure 9.3 shows these axes relative to the fence, with the positive normal axis ($+N$) in the opposite direction to the impact direction, and the positive tangential axis ($+T$) in the downward direction. This means that the normal component of the impact velocity, which is toward the net, is negative ($-v_i$). The orientations of the impact (i) and restitution (f) velocity vectors—θ_i and θ_f—are defined relative to the fence inclination.

The insets on Figure 9.3 show the usual direction of rotation (clockwise) for rock falls due to the frictional forces acting at the contact point with the slope (see Section 4.5). Since the direction of rotation at the contact point is in the direction of the positive tangential axis, the clockwise rotation is defined as a positive rotational velocity. The combined effect of the rotation and normal impulse p_N is to generate a frictional force ($\mu \cdot p_N$) at the contact point that modifies the shear component of the velocity and the restitution angle as discussed in Section 4.5.

Worked Example 9A below shows the method of calculating the restitution velocities of a rock fall impacting a fence at an oblique angle using Equations (4.20) to (4.24) that apply impact mechanics to model velocity changes during impact.

Worked Example 9A—velocity changes during oblique impact: for oblique impact of a body with a fence at the orientation shown in Figure 9.3(b). The values for the impact parameters are

$$V_i = 22 \ \text{m} \cdot \text{s}^{-1}; \ \omega_i = 18 \ \text{rad} \cdot \text{s}^{-1}; \ \theta_i = 50°$$

The body is assumed to be a cube with side length of 0.6 m (see Section 2.1 on Ehime test site, Japan), the properties of which are

mass, m = 520 kg; radius (diagonal) = 0.43 m; radius of gyration, k = 0.35 m

For this impact geometry, the impact normal (N) and tangential (T) velocity components are as follows:

$$V_{iT} = V_i \cos \theta_i = 22 \cos 50 = 14.1 \text{ m} \cdot \text{s}^{-1}$$

and

$$V_{iN} = -V_i \sin \theta_i = -22 \sin 50 = -16.9 \text{ m} \cdot \text{s}^{-1}$$

The final tangential velocity component is calculated using Equation (4.20):

$$V_{fT} = V_{iT} - \frac{(V_{iT} + r \cdot \omega_i)}{(1 + r^2/k^2)} \tag{4.20}$$

$$= 9.9 \text{ m} \cdot \text{s}^{-1}$$

The final normal velocity is calculated using the normal coefficient of restitution, e_N.

$$V_{fN} = -V_{iN} \cdot e_N = -(-16.9 \cdot 0.35) = 5.9 \text{m} \cdot \text{s}^{-1} \tag{4.21}$$

where e_N is defined by the impact angle θ_i as demonstrated in Section 5.2.2 and Figure 5.7. Equation (5.4) approximately relates e_N and θ_i as follows:

$$e_N = 19.5 \ \theta_i^{-1.03} \tag{5.4}$$

Therefore, for an impact angle of 50°, the approximate normal coefficient of restitution is 0.35. The final rotational velocity is given by Equation (4.22):

$$\omega_f = \omega_i - \frac{r}{k^2} \frac{(V_{iT} + r \cdot \omega_i)}{(1 + r^2/k^2)} \tag{4.22}$$

$$= -11.6 \text{rad} \cdot \text{s}^{-1}$$

The final restitution velocity v_f and angle θ_f can be calculated from the final tangential and normal velocities components as follows:

$$V_f = \sqrt{V_{fT}^2 + V_{fN}^2} = 11.5 \text{ m} \cdot \text{s}^{-1} \tag{4.23}$$

and

$$\theta_f = \text{atan}\left(\frac{V_{fN}}{V_{fT}}\right) \tag{4.24}$$

$$= 31 \text{ degrees}$$

In Figure 9.3(b), the velocity vectors are drawn to scale to show the results of the impact calculations. These calculations show that the redirection of the body by the net results in the velocity being reduced by 48%, from $V_i = 22$ m · s^{-1} to $V_f = 11.5$ m · s^{-1}. The velocity will decrease up to the point of maximum deflection and will not be zero at any time.

The change in the rotational velocity from $\omega_i = +18$ rad · s^{-1} to $\omega_f = -11.6$ rad · s^{-1} indicates that the direction of rotation reverses during impact. Model tests of bodies impacting the net at an oblique angle indicate the reversal of the rotation direction.

9.2.2 Energy Changes during Impact with a Fence

The velocity changes during impact with the fence discussed in Section 9.2.1 above and shown in Figure 9.3 can also be used to calculate the energy changes during impact and determine how much of the energy is absorbed by the net.

For the fence configuration shown in Figure 9.3a) where the impact is normal to the net, the velocity of the rock is reduced from the initial velocity v_i at the moment of impact to zero at the time of maximum deflection. Therefore, the total impact energy of the rock fall is absorbed by the net, i.e. ($E = \frac{1}{2} m \cdot v_i^2$). This type of impact is also illustrated in Figure 9.2a) where the rocks that have been stopped by the net accumulate in the base of the fence.

For the fence configurations shown in Figures 9.3b) and 9.4 where the impact is oblique to the net, the velocity decreases during impact, but the rock is not stopped. Therefore, only a portion of the impact energy is absorbed by the fence, with the remainder of the energy being retained by the moving rock.

Worked Example 9B below illustrates the calculation method for an oblique impact with a fence.

Worked Example 9B—energy changes during oblique impact: calculations of the velocity changes that occur during an oblique impact are described in Worked Example 9A. The corresponding energy changes during impact are calculated as follows.

For an oblique impact of 50° with the net, the velocity decreases from $22\ \text{m} \cdot \text{s}^{-1}$ to $11.5\ \text{m} \cdot \text{s}^{-1}$, and the rotational velocity changes during impact from $18\ \text{rad} \cdot \text{s}^{-1}$ to $-11.6\ \text{rad} \cdot \text{s}^{-1}$.

The moment of inertia of the cubic body, $I = m \cdot k^2 = 520 \cdot 0.35^2 = 63.7\ \text{kg} \cdot \text{m}^2$.

Assuming that the mass of the rock remains constant during impact, the initial and final energies are as follows:

Initial energy: $KE_i = \frac{1}{2} m \cdot V_i^2 = \frac{1}{2}\ 520 \cdot 22^2 = 125.8\ \text{kJ}$; $RE_i = \frac{1}{2} I \cdot \omega_i^2 = \frac{1}{2}\ 63.7 \cdot 18^2 = 10.3\ \text{kJ}$
Total impact energy, $E_i = (125.8 + 10.3) = 136.1\ \text{kJ}$

Final energy: $KE_f = \frac{1}{2} m \cdot V_f^2 = \frac{1}{2}\ 520 \cdot 11.5^2 = 34.4\ \text{kJ}$; $RE_f = \frac{1}{2} I \cdot \omega_f^2 = \frac{1}{2}\ 63.7 \cdot (-11.6)^2 = 4.3\ \text{kJ}$
Total final energy, $E_f = (34.4 + 4.3) = 38.7\ \text{kJ}$

These calculations show that, for an oblique impact, the impact energy that is absorbed by the net is 97.4 kJ (136.1–38.7), which is only 72% of the impact energy compared to 100% of the impact energy for a normal impact. (See Figure 9.3(b).)

9.2.3 Energy Efficiency of Fences

The discussion on the velocity and energy changes related to the impact configuration in Sections 9.2.1 and 9.2.2 above can be used to quantify the relationship between energy absorption by the fence and the impact configuration.

For a normal impact where the rock fall is stopped by the net, all the impact energy is absorbed by the structure, which has to be designed to withstand the full impact energy. In contrast, for an oblique impact only a portion of the impact energy is absorbed by the structure. The portion of the impact energy that is deflected by the net can be expressed in terms of the efficiency of the structure as follows:

$$Efficiency,\ E_e = \frac{Restitution\ energy,\ E_f}{Impact\ energy,\ E_i}.100\% \qquad (9.1)$$

For the two fence configurations shown in Figures 9.3(a)(b), the efficiencies are, respectively, 0% for the normal impact and $E_e = 39.2/138.8 = 28\%$ for the oblique impact where the rock fall is redirected by the net.

The benefit of a high-efficiency fence is that, for the same impact energy, it can be constructed from lighter weight materials than a zero efficiency fence. Conversely, design impact loads will be greater with increasing energy efficiency of the fence. The influence of fence configuration on energy efficiency is discussed further in Section 9.3 and 9.4 of this chapter.

9.2.4 Configuration of Redirection Structures

Figure 9.3 shows how an oblique impact on a net can be achieved by inclining the fence upslope. The same impact configuration occurs for the hanging net shown in Figure 9.4 where a vertical hanging net, unconstrained along its lower edge, is suspended from a series of posts attached to the rock face. When rock falls impact the net, their velocity is reduced, and they are redirected into a catchment area at the base of the net; the rock falls that accumulate in the catchment area can be readily removed by maintenance equipment. In comparison, for the net shown Figure 9.2(a) removal of accumulated rock falls requires access to the slope to detach the lower edge of the net from the support cables.

The hanging net configuration shown in Figure 9.4 is applicable at locations where the rock face is near vertical and the ditch width is limited so that it is not possible to construct the type of fence shown in Figure 9.2(a). Construction details of hanging nets are described in more detail in Section 10.5.2.

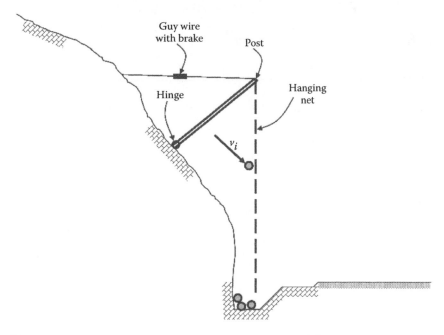

Figure 9.4 Hanging net installed on steep rock face that redirects rock falls into containment area at base of slope, where cleanout of accumulated rock falls is readily achieved.

9.2.5 Hinges and Guy Wires

Both the fence configurations shown in Figures 9.3(b) and 9.4 incorporate hinges at the base of the posts and guy wires, equipped with brakes, to hold the posts in place. These two components of the fences are energy-absorbing features that, together with the net itself, ensure that the entire structure is flexible. Ideally, all components of the fence are equally flexible because forces tend to concentrate in the stiffest part of the structure. The absorption of impact energy with time in rock fall containment structures is discussed in Section 9.3, with the design objective being to absorb energy uniformly during impact in order to minimize the forces induced in the structure.

Commercially available braking elements in the guy wires, comprising a variety of loops, coils or sliding connections, that absorb energy by plastic deformation and need to be replaced when the impact energy exceeds the Service limit state (see Figure 8.2). Nets are generally designed for easy replacement of brakes.

Many commercial rock fall fences also incorporate hinges at the base of posts, the primary function of which is to allow the post to deflect in the event of an impact energy that exceeds the Service limit states energy. When this occurs, the post will deflect without damaging the foundation so that repairs to the fence are limited to restanding the posts and replacing any damaged net and brake components; this is less costly than rebuilding the foundations. Figure 9.5 shows a post for a hanging net that was impacted by a snow avalanche, with an energy that exceeded the service limit state but was less than the ultimate limit state. For this impact, the net and posts were deflected by several meters and the brakes on the guy wires were activated and needed to be replaced, but no damage occurred to any net component.

Figure 9.5 Low-friction hinge at base of post allowed the post to deflect, with no damage to the foundation, during an impact that exceeded the service limit states energy (above Shuswap Lake, British Columbia, Canada).

In contrast to the energy-absorbing features of brakes, most hinges are low-friction units that do not absorb energy when the post deflects. It is believed that inclusion of frictional elements into hinges will allow the hinges to both deflect in order to prevent damage to the foundation and to absorb energy. The combined activation of the guy wire brakes and the postfrictional hinges will produce a reasonably uniform absorption of energy during the time of impact. As discussed in Section 9.3, uniform energy absorption with time minimizes the forces that are induced in the fence by the impact.

9.3 MINIMIZING FORCES IN ROCK FALL PROTECTION FENCES

A fundamental design feature of effective rock fall fences and barriers is the flexibility of the structure so that energy is absorbed in the structure during impact. An example of a flexible structure is a net fabricated with woven steel wires that is supported on posts with flexible hinges and deformable guy wires; most components of the net will deform and absorb energy with impact (Figure 9.2(a)). This energy absorption will not, up to the design (service) energy, cause any damage to the structure. In contrast, rigid structures such as mass concrete walls have essentially no capacity to absorb energy except by fracturing, resulting in permanent damage (Figure 9.2(b)).

9.3.1 Time–Force Behavior of Rigid, Flexible, and Stiff Structures

The capacity of a structure to absorb energy can be quantified by examining the force generated in the structure during the time of impact. This behavior is illustrated in Figure 9.6 by [time–force] plots for three different types of structure. These plots do not show the behavior of actual structures, but demonstrate the difference between the rigid, flexible, and stiff structures, the characteristics of which are discussed below.

The objective of studying [time–force] behavior is to design structures in which the force induced by the impact in the structure is minimized. That it is preferable that structures that

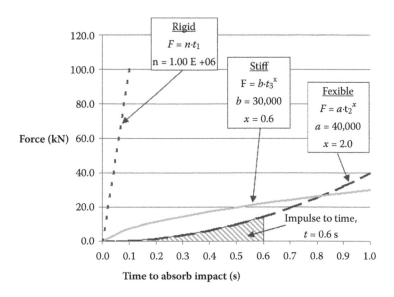

Figure 9.6 Relationship between time of impact and force generated in rigid, flexible, and stiff fences.

have high-energy efficiency as defined in Section 9.2.3. By minimizing the induced force, it is possible to construct fences that use lighter-weight materials, and are less costly, than if the full impact energy has to be absorbed by the fence.

Figure 9.6 shows typical [time–force] relationships for rigid, flexible, and stiff fences, and the equations that define these relationships. The physical characteristics of these three types of structures are discussed below.

1. **Rigid structures**—For a rigid structure such as a mass concrete wall, the primary means of energy dissipation is shattering of the brittle concrete and permanent damage to the structure (Figure 9.2(b)). This behavior is illustrated in Figure 9.6 as a straight line with a steep gradient representing the very short time of impact before fracturing occurs at a force equal to the strength of the concrete. This behavior shows that concrete walls are not effective rock fall barriers.

2. **Flexible structures**—A variety of rock fall fences are available in which all the components are flexible, and energy is absorbed by deformation of these components without damage (Figure 9.2(a)). For these structures in which the components are highly flexible, a significant amount of deformation has to occur before energy starts to be absorbed. Furthermore, if the bases of the support posts incorporate low-friction hinges, no energy is absorbed as the posts rotate on the hinges. This behavior is illustrated in Figure 9.6 by a curve that shows little force generated in the fence during the early time of impact, after which the force increases with increasing deformation of the net and supports. Flexible structures are effective in containing rock falls but must be strong enough to absorb the full impact energy since the rock falls are stopped by the net.

3. **Stiff structures**—Stiff structures are modified flexible fences that are designed to start absorbing energy from the moment of impact. The behavior is illustrated in Figure 9.6 as a curve with an approximately constant gradient throughout the time of impact, after the initial contact. This [time–force] behavior is achieved by the following characteristics of fence design:

   • *Attenuation and redirection*—Section 9.2.1 above and Figure 9.3 demonstrate the difference between normal and oblique impact of a rock fall with a fence, and how the energy efficiency can be improved by the rock being redirected rather than stopped by the structure.

   • *Flexible nets*—A variety of nets are available in which the steel wire or cable components deform during impact. Nets that are slightly stretched between the support cables will be somewhat stiffer than nets that are very loosely attached to the cables.

   • *Frictional hinges*—Posts supporting rock fall nets are usually anchored to the ground with rock or soil anchors, or attached to concrete blocks buried in the ground. The connection between the posts and the foundation can be rigid or hinged, with the advantage of hinged connections being that the post can deflect without damage to the foundation, resulting in reduced maintenance costs. However, if the hinge is very flexible, no energy is absorbed during deflection. In a stiff structure, hinges are frictional so that the posts can deflect without damaging the foundations but energy will be absorbed during deflection.

   • *Energy-absorbing guy wires*—It is common that the guy wires supporting fence posts incorporate brakes that deform plastically and absorb energy where the impact energy exceeds a design threshold.

If these four features have the same [time–force] and stiffness characteristics then they will absorb energy simultaneously, starting at the time of first contact. These features and the performance of stiff fences are discussed in more detail in Section 9.5, Model testing.

9.3.2 Energy Absorption by Rigid, Flexible, and Stiff Structures

The [time (t)–force (F)] plots in Figure 9.6 are modeled as a straight line for the rigid concrete wall, and as power curves for the flexible and stiff fences according to the following relationships:

- Rigid structure:

$$F = n \, t_1 \tag{9.2}$$

where n is the gradient of the line analogous to the stiffness of the structure and t_1 is the duration of contact. For a mass concrete, unreinforced wall an assumed value for the gradient is 1E6 units [MLT^{-3}], consistent with the short duration of contact and the large force generated in the structure in this time.

- Flexible structure:

$$F = a \, t_2^x \tag{9.3}$$

where a is a constant, t_2 is the duration of contact, and x is an exponent that has a value greater than 1; an exponent greater than 1 generates a curve for which the force is negligible during the initial contact and then increases rapidly with time. For this illustration of the behavior of a flexible structure, the value of a is 40,000 and the exponent x is 2.0.

- Stiff structure:

$$F = b \, t_3^x \tag{9.4}$$

where b is a constant, t_3 is the duration of contact, and x is an exponent that has a value less than 1; an exponent less than 1 generates a curve for which the force increases steadily throughout the contact time. For this illustration of the behavior of a stiff structure, the value of b is 30,000 and the exponent x is 0.6.

For [time–force] plots illustrated in Figure 9.6, the area under the curve equals the impulse of a rock fall absorbed by the structure over any specified time interval (see Section 4.2 and Equation (4.1)). The impulse (or momentum) of the rock fall at the time of contact with the structure is equal to ($m \cdot v_i$), where m is the mass of the rock fall and v_i is the impact velocity. The duration of the contact time must be sufficient for the impact impulse to be absorbed, and this duration will differ with the stiffness of the structure. That is, for a rigid structure, the contact duration, t_1 will be short because of the rapid increase in the force, while for flexible and stiff structures the contact duration t_2 and t_3 will be longer than t_1 as the force increases slowly during contact. The objective of this design approach is to minimize the force in the structure, and this is achieved with a relatively long duration contact time.

The impulse absorbed by the structure can be obtained by integration of the [time–force] plots to find the areas under the curves over the duration of contact, as follows:

- **Rigid structure**—From Equation (9.2), impulse absorbed by the structure over contact time t_1 is

$$p = \int_0^{t_1} F \, dt = n \int_0^{t_1} t \, dt = \frac{1}{2} n \cdot t_1^2 \tag{9.5}$$

- **Flexible structure**—From Equation (9.3), impulse absorbed by the structure over contact time t_2, for $x > 1$ is

$$p = \int_0^{t_2} F \, dt = a \int_0^{t_2} t^x \, dt = \frac{a}{(x+1)} t_2^{(x+1)} \tag{9.6}$$

- **Stiff structure**—From Equation (9.4), impulse absorbed by the structure over contact time t_3, for $x < 1$ is

$$p = \int_0^{t_3} F \, dt = b \int_0^{t_3} t^x \, dt = \frac{b}{(x+1)} t_3^{(x+1)} \tag{9.7}$$

In Figure 9.7, these integrated equations are plotted to show the relationship between the time of contact and the impulse absorbed by the fence.

If the impulse of the rock fall is defined by the mass and velocity at the time of contact, then Equations (9.5), (9.6), and (9.7) can be solved to find the duration of contact for each type of structure. Referring to the Worked Example 9A above, for a rock fall mass, m of 520 kg (1,150 lb) and impact velocity, v_i of 22 m · s^{-1} (70 ft · s^{-1}), the impulse is 11,440 kg · m · s^{-1}. For this impulse, the corresponding contact durations are:

- **Rigid structure**: For n = 1E6, t_1 = 0.15 seconds.
- **Flexible structure**: For a = 4E4 and x = 2.0, t_2 = 0.95 seconds.
- **Stiff structure**: For b = 3E4 and x = 0.6, t_3 = 0.73 seconds.

These contact durations can be substituted in Equations (9.2), (9.3), and (9.4) to find the maximum forces generated in the structures during contact as follows:

- **Rigid structure**: For t_1 = 0.15 seconds, F = 150 kN.
- **Flexible structure**: For t_2 = 0.95 seconds, F = 36.1 kN.
- **Stiff structure**: For t_3 = 0.73 seconds, F = 25.0 kN.

Figure 9.7 shows the [time–impulse] plots for the three types of structure that are given by Equations (9.5), (9.6), and (9.7) and represent the areas under the [time–force] plots given in Figure 9.6. Figure 9.8 shows the force generated in the structure due to the absorption of the applied impulse. The plots in Figures 9.7 and 9.8 and the values of the maximum forces show that stiff structures absorb impact energy more efficiently than both rigid and highly flexible structures. For the example presented in this section, the maximum force generated in a stiff structure is only 15% (22.0/151 = 0.15) of that for a rigid structure, and 61% (22.0/36.1 = 0.61) of that for a flexible structure.

The calculated forces in the three structures can be illustrated graphically on the [time–force] plot in Figure 9.8.

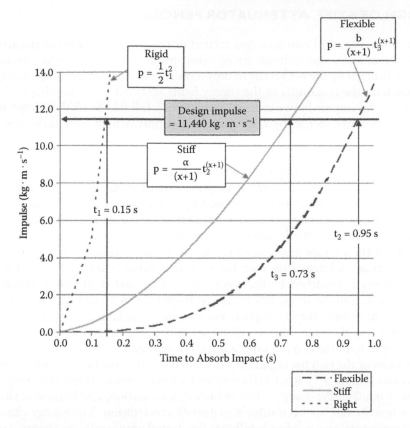

Figure 9.7 Plot of duration of impact against rock fall impulse absorbed by fence; curves are developed by integrating [time–force] equations shown in Figure 9.6 to give the area under [time–force] curves.

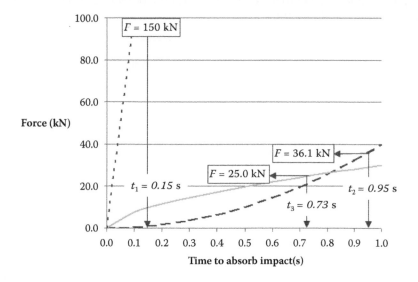

Figure 9.8 [Time–force] plot for rigid, flexible, and stiff structures showing force generated in structure at duration taken to absorb impact impulse.

9.4 DESIGN OF STIFF, ATTENUATOR FENCES

This chapter has discussed two concepts related to the energy absorption efficiency of rock fall fences: attenuation and uniform energy absorption. First, an oblique impact in which the rock is redirected by the net results in only a portion of the impact energy being absorbed by the fence, with the remainder of the energy being retained in the moving rock. This is in contrast to normal impact between the net and a rock fall where all the energy is absorbed as the rock is brought to rest by the fence. Fences that deflect rock falls may be termed "attenuator" structures.

Second, for a fence designed so that the energy is absorbed uniformly with time, the forces generated in the fence by the impact will be less than that for both rigid and highly flexible structures. Fences with uniform energy absorption with time may be termed "stiff" structures.

Figure 9.3(b) shows a fence that is inclined upslope, while Figure 9.9 shows an alternative configuration in which the posts supporting the net are made of two segments; in both cases, a hinge is located at the connection between the base of the post and the foundation. The function of the upslope segment of the net is to create impacts at an oblique angle so that the net acts as an attenuator, with the rock falls being redirected so only a portion of the impact energy is absorbed by the fence. Also, if the net is slightly tensioned between the support posts, and the hinge has frictional resistance to movement, energy is absorbed approximately uniformly during impact. For the fence configuration shown in Figure 9.9, rock falls impacting the lower (downsloping) segment of the net will deform the net and then impact the slope, with the ground taking out the remaining energy.

The functions of the two net segments are related to the trajectories of rock falls and their related energies (Figure 9.9). Rock fall energies are lowest immediately after impact with the ground where much of the energy is lost in plastic deformation, and highest at the end of the trajectory when the fall has been subject to gravity acceleration. The energy changes during impact and trajectory phases of rock falls are illustrated graphically on energy partition diagrams (Figures 6.5 and 6.7), on which the low-energy portions of the fall after each impact can be clearly identified. Based on the principles shown in the energy partition diagrams, fences should be designed to withstand the high-energy portion of falls, that is, at the end of trajectories, if it is not possible to locate the fence on the outside of a bench where rock falls will impact as shown in Figure 9.1.

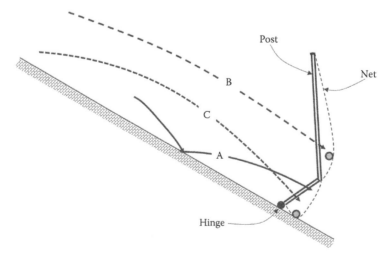

Figure 9.9 Rock fall trajectories impacting attenuator-type fence. Trajectory A—low-energy impact close to slope impact point; trajectories B and C—high-energy impacts distant from prior impact points.

Figure 9.9 shows three trajectories: trajectory A is a low-energy impact occurring soon after impact, while trajectories B and C are higher-energy impacts occurring a longer time after impact. In Figure 9.9, trajectory B impacts the upper portion of the fence at an oblique angle and is redirected toward the ground by the net so the energy is attenuated. Trajectory C impacts the lower part of the fence such that the net is deflected and impacts the ground where the impact energy is absorbed.

The fence configuration shown in Figure 9.9 is a concept that illustrates the principles of efficient fence design that may have application in commercial products.*

9.5 MODEL TESTING OF PROTECTION STRUCTURES

This chapter discusses both attenuation of rock falls by rock fall fences, and the performance of stiff structures in terms of impact mechanics. In order to validate these concepts and calculations, model tests of an attenuator fence were carried out as described below.

9.5.1 Model-Testing Procedure

The model-testing procedure involved using a baseball pitching machine to propel baseball-size projectiles at a 1/20 scale model of a rock fall fence. It was possible to closely control the velocity of the projectiles, and to have both rotating and nonrotating impact. Rotation of the projectile was achieved by having the projectile impact the slope just above fence so that the direction of rotation was counterclockwise, which is the same direction as usually occurs in actual rock falls. Impact of nonspinning particles was achieved by having the projectiles impact the net without first impacting the slope.

The model tests were carried out on the fence with the configuration shown in Figure 9.10 that is designed to have both "attenuation" and "stiff" force-deformation characteristics as discussed in this chapter.

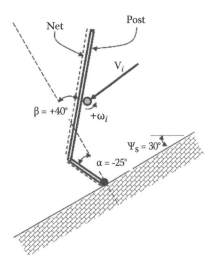

Figure 9.10 Configuration of rock fall fence used in model tests. Orientations of hinged posts are defined by angles α and β, measured relative to the normal to the slope.

---

* Energy Absorbing Barrier, United States and European patents pending.

The orientations of the two post segments are defined by the angles α and β are measured relative to the normal to the slope, with upslope angle being positive. For the tests, the lower post was fixed at an orientation of α = –25°, and the orientation of the upper post was varied between β = –25° for normal impacts, and β = +60° for oblique impacts. For each orientation, rotating and nonrotating impacts were tested. The objective of the tests was to find a relationship between the orientation of the upper net and the energy absorption efficiency, see Equation (9.1).

9.5.2 Model Test Parameters

The baseball pitching machine allowed the translational velocity of the projectile, as well as its rotational velocity to be controlled, with values selected to achieve impacts with high enough energy to significantly deflect the net. The translational velocities varied between 7 and 27 m · s^{-1} (25 to 90 ft · s^{-1}), and the rotation was counterclockwise for the model shown in Figure 9.11.

The projectile was a dense rubber ball with a dimpled surface, a diameter of 75 mm (3 in) and a mass of 140 gram (0.3 lb). For a sphere, the rotation parameters are: radius of gyration, $k = (2/5)^{0.5} \cdot r = 0.024$ m and moment of inertia, $I = m \cdot k^2 = 7.9\text{E-}5$ kg · m^2.

9.5.3 Results of Model Tests

For the model tests, the image analysis software, ProAnalyst© (Xcitex, 2008), was used to find the impact (initial) velocities, V_i, ω_i and final (restitution) velocities, V_f, ω_f from which the initial and final total energies could be calculated. Figure 9.11 shows a typical result of an oblique angle impact between the projectile and the fence inclined upslope, and the

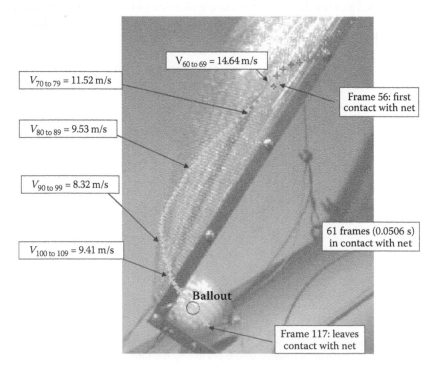

Figure 9.11 Path of deflected projectile after impact with net-oriented upslope (β = +60°). Approximate velocities at 10 frame intervals (0.0083 s) during impact shown.

reduction in velocity that occurs during impact. The image shows the redirection by the net of the body, which "rolls" down the net during impact; contact duration is 0.0506 seconds (61 frames). During this contact duration the velocity decreases from $V_i = 14.6$ m · s⁻¹ (48 ft · s⁻¹) at the point of impact to a minimum velocity of 8.3 m · s⁻¹ (27 ft · s⁻¹) between frames #90 to #99, approximately 0.03 seconds after impact. After this time, the velocity increases slightly to $V_f = 9.41$ m · s⁻¹ (31 ft · s⁻¹) as projectile starts to rebound off the net. These velocity changes illustrate "attenuation" type behavior of the fence.

The changes in velocity were then used to find a relationship between the fence orientation defined by the angle β and the energy efficiency E_e (see Equation [9.1]), where the energy efficiency increases when more of the impact energy retained by the moving projectile and less is absorbed by the net. For a projectile that is stopped by the net and all the impact energy is absorbed, the energy efficiency is zero.

Figure 9.12 shows the relationship between the angle β and the energy efficiency. That is, when normal impact occurs (β = 0), all the energy is absorbed and $E_e = 0$. For a nonrotating body, the energy efficiency increases as the upper net is inclined further upslope (angle β increases) showing the effect of the body being redirected due to the oblique impact angle. When β = 60°, only about 50% of the impact energy is absorbed.

The test results also provided information on the changes in velocity and impulse over the duration of the impact. For example, the relationship between the velocity and duration that the body is in contact with the net is shown in Figure 9.13 for a variety of tests in which the β angle varied from 0 to 60°. In Figure 9.13, the velocity decreases during contact from an initial value of about 15 m · s⁻¹ to a minimum value of about 7 m · s⁻¹ before increasing slightly as the body starts to rebound off the net. At no time does the velocity approach zero indicating that only a portion of the impact energy is absorbed by the net.

The velocity, V data shown in Figure 9.14 can be expressed as an impulse ($p = m \cdot V$) where m is the mass of the body. For the baseball-size body used in the tests with a mass of 0.14 kg (0.3 lb), impulse values were calculated for the corresponding velocities measured during the impact. Figure 9.14 shows the relationship between the duration of the impact and the percentage of the impact impulse that is absorbed by the fence during impact. That

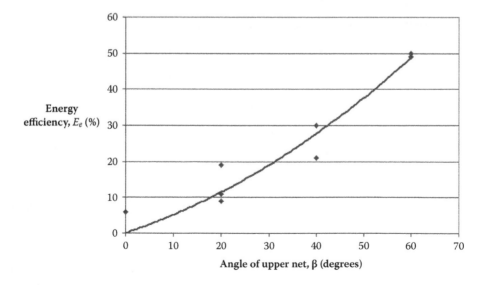

Figure 9.12 Relationship between energy efficiency and angle β of upper net, for a nonrotating body.

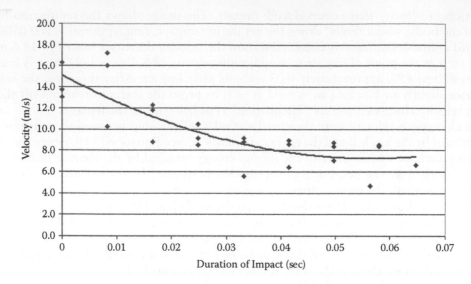

Figure 9.13 Relationship between the duration of impact and change in velocity during impact.

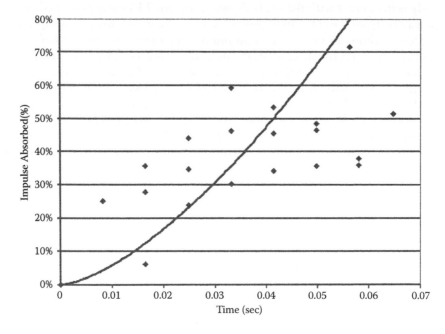

Figure 9.14 Relationship between duration of impact and the amount of impulse that is absorbed by the fence; compare with Figure 9.7 for stiff structures.

is, the impulse begins to be absorbed very soon after impact and increases steadily during the impact.

The form of the actual [time–impulse] relationship shown in Figure 9.14 can be compared with the theoretical relationship between [time–force] shown in Figure 9.6 where the behavior of rigid, flexible, and stiff structures are compared. It is apparent that the fence configuration used for the model tests has produced a stiff structure that absorbs energy throughout the impact, thus minimizing the induced force.

This chapter has examined the theory of designing efficient protection structures that redirect (attenuate) rather than stop rock falls. This concept has been used in the construction of hanging net-type fences as shown in Figure 9.4, an example of which is shown in Figure 10.20 where it is providing rock fall protection for a railway. The author has been involved with the design and construction of about 25 hanging nets that have been impacted by many hundreds of rock falls, almost all of which have caused no significant damage to the fence. This performance, together with theory and model tests, would indicate that attenuator-type structures are effective in containing rock falls.

Chapter 10

Rock Fall Protection I—
Barriers, Nets, and Fences

This chapter describes a variety of commonly used rock fall protection structures, ranging from inexpensive ditches to high-energy-capacity wire-rope fences. Design principles for reinforced concrete rock fall sheds and wire-rope canopies are discussed in Chapter 11.

The selection of a protection structure that is suitable for a particular site will depend on the following three conditions:

a. **Impact kinetic energy**—The mass, m, and velocity, V, of the rock that may impact the structure defines the impact kinetic energy ($KE = \frac{1}{2} m \cdot V^2$). The mass of the rock fall depends on the discontinuity spacing and the dimensions of the blocks that can be formed. The rock strength also influences the impact energy because blocks of weak rock tend to break into smaller fragments as they impact the slope, while blocks of strong rock retain their mass during the fall. The velocity of the fall depends on the fall height, the slope angle, and the effective friction coefficient of the slope surface as defined by Equation (3.13). The mass and velocity data are then combined to select the service and ultimate limit states design energies (see Section 8.2).

   If appropriate, the rotational energy ($RE = \frac{1}{2} I \cdot \omega^2$) can be included in the total impact energy. For example, Worked Examples 6B and 6C discuss the energy changes at impact point #A26 at the Tornado Mountain rock fall site where the rotational energy is about 18% of the total impact energy.

b. **Site geometry**—The main slope geometry parameters relevant to the design of protection structures are the fall height and the slope angle. The site geometry also dictates the trajectory path and the required height of the structure. The distance between the base of the slope and the facility to be protected dictates the space available to construct protection measures. It is usually preferable to locate the protection structure such as a barrier or fence at the same level as the road or railway, for example, since this facilitates both construction and maintenance. Where the available space at the base of the slope is limited or cannot be created by excavating the base of the slope, it will be necessary that the protection structure, such as an anchored fence, be located on the slope.

   One factor to consider in the fence location is the deflection that will occur during impact, with sufficient clearance being provided that the house, for example, is not struck when the fence is deformed by the impact.

c. **Cost-benefits**—The construction cost of a protection structure should be consistent with the expected cost of a rock fall damaging the facility below the slope. As discussed in Section 8.5.5 on decision analysis, protection options can be compared in terms of their total expected cost, which is the sum of the construction cost of the protection structure, and the expected cost of a rock fall that may include delays to traffic, damage to equipment and possible injuries, and required slope stabilization. Expected costs are the product of the cost of an event and the probability of its occurrence. This probabilistic approach accounts for the uncertainty in the size and frequency of rock fall events and their consequences. In summary, a costly protection structure such as a reinforced concrete shed is usually only justified where large falls are possible and the consequences are severe, such as disruption to a high-speed railway or high-traffic-volume highway.

Figure 10.1 shows a variety of protection structures and ranges of their impact energy capacity. With respect to their approximate construction cost, ditches are usually the least expensive and concrete rock shed the most expensive.

The protection measures described in this chapter are divided into two categories as follows:

- **Ditches and barriers**—These structures, which are usually constructed at the base of the slope where space permits, are often a reliable, low-cost, and low-maintenance protection measure.
- **Net and fences**—These structures, which can be constructed either on the slope or at the base of the slope, require reliable information on impact energies and trajectories in order to prepare dependable designs.

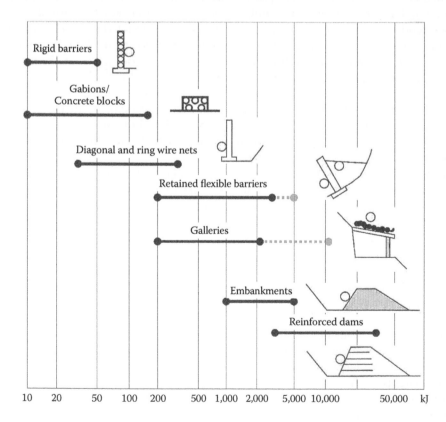

Figure 10.1 Ranges of impact energy capacity for a variety of protection structures. (After Vogel, T. et al. [2009]. Rock fall protection as an integral task. *Structural Engineering International*, SEI Vol. 19[3], 304–12, IABSE, Zurich, Switzerland, www.iabse.org.)

10.1 DITCHES AND BARRIERS

Ditches and barriers are often a cost-effective rock fall protection measure, and well-proven design methods are available to determine the required dimensions. A variety of structures are discussed below.

10.1.1 Ditch Design Charts

Where space is available between the base of the slope and the at-risk structure such as a house or highway, a ditch will often provide reliable protection. Furthermore, construction and maintenance costs are low because they can be can be located beside the structure, with no need to access the slope.

The required dimensions of a ditch, defined by its base width, inclination, and depth, are related to the height of the rock face, and its slope angle (Figure 10.2). The ditch design chart in Figure 10.2 was developed from field observations and shows the following behavior of falling rock (Ritchie, 1963). With increasing height, the fall velocity and the trajectory height increase, requiring a corresponding increase in the ditch dimensions. With respect to the slope angle, at an angle steeper than about 75° to 80°, rocks fall close to the face with shallow trajectories and land near the base of the slope; such that a narrow ditch is usually adequate. However, at slopes angles between about 55° and 75°, rocks will have higher trajectories and rotation, and land farther from the base of the slope, requiring a wider ditch. At slope angles flatter than about 55°, rocks tend to roll down the slope with low trajectories and the required ditch dimensions are reduced.

Generally, the flattest slope on which rock falls will occur is the angle of repose of loose rock fragments of about 37°, which is the angle at which talus slopes form (see Figure 1.1). However, Figure 1.1 as well as the Tornado Mountain case study (see Section 2.2.2) show that rock falls can move on slopes as flat as 20° if the slope surface is smooth and the rock is larger and has more energy than previous falls that have accumulated on the talus. The farthest distance that rocks may travel past the base of a talus slope is defined by the angle of about 27.5° (below the horizontal) measured from the top of the talus slope (Hungr and Evans, 1988).

As an update to the ditch design chart developed by Ritchie shown in Figure 10.2, an extensive rock fall testing program was carried out in Oregon in which 11,250 rocks were rolled down rock faces with angles ranging from vertical to 45°, and heights ranging from 8 to 24 m (25 to 80 ft) (Pierson et al., 2001). One of the test results is discussed in Section 2.1. The primary purpose of these tests was to determine the required ditch dimensions for highways where a uniformly sloping "recovery zone" is required between the edge of pavement and the base of the rock cut; the recovery zone allows drivers who inadvertently leave the traveled surface to recover without impacting a barrier or falling into a ditch. To meet this requirement, the ditches in the tests were defined by their slope angles of 4H:1V (14°), 6H:1V (9.5°), and horizontal, and the tests recorded the location of the first impact and the distance that the rocks rolled across the recovery zone.

Figure 10.3 shows an example of a design chart for a 15 m (50 ft) high cut at a face angle of 76° (0.25H:1V). The plots of the test results show the relationship between the catchment width to contain the falls and the percentage of rocks retained within this width. For example, for a catchment zone sloping at 6H:1V, 70% of the falls stop within a width of 3.4 m (11 ft), while 95% of the falls stop within a distance of 6 m (20 ft). In addition, the results show the significant benefit of a sloped catchment area with 90% of the falls rolling to 15.5 m (51 ft) on a horizontal catchment area, while they only roll 6.4 m (21 ft) for a catchment area sloped at 4H:1V.

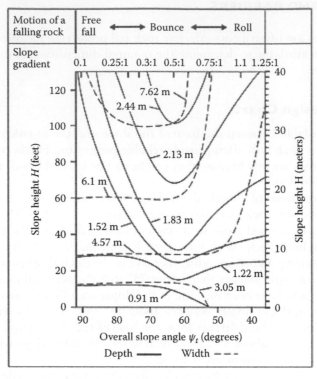

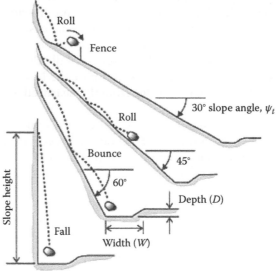

Figure 10.2 Ditch design chart showing the required width and depth of the ditch in relation to the height and angle of the slope. (From Ritchie, A. M. [1963]. An evaluation of rock fall and its control. *Highway Research Record* 17, Highway Research Board, NRC, Washington DC, pp. 13–28.)

The value of the results being expressed in terms of percent retained is that the ditch dimensions can be designed to suit the level of reliability of the protection (see Section 8.3). That is, for a frequently occupied building it would be appropriate that the width of the catchment area be sufficient to contain 100% of the falls. However, for a low-traffic-volume road, it may be acceptable to contain only 80% of the falls.

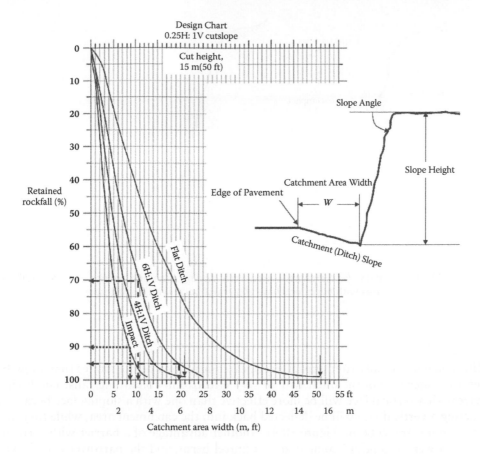

Figure 10.3 Ditch design chart for a 15 m (50 ft) high cut at a face angle of 0.25H:1V showing relationship between the catchment area width and the percent of rock falls retained (Pierson et al. 2001).

A useful set of data from the Oregon tests is the first impact point in the ditch. For the data shown in Figure 10.3, 90% of the falls first impact the catchment area within 2.4 m (8 ft) of the base of the face. However, once the rocks impact the catchment area, they then roll for a considerable distance toward the highway, for a distance of as much as 16 m (51 ft) from the base of the cut, in the case of the horizontal catchment area. These roll-out distances are partly generated by the rotation of the block as it impacts the face during the fall, and then rolls along the ground.

10.1.2 Ditch Geometry

Figure 10.3 shows the significant influence of ditch geometry on the "run out" distance that blocks of rock bounce and roll across a catchment area. The path of falls across the catchment area can be ascertained from the relationship between the impact angle θ_i and the normal coefficient of restitution e_N, where e_N has values of about 0.1 to 0.2 for steep impact angles (see Figure 5.5). For rock falls on steep faces that impact a horizontal catchment area, the incidence angle will be close to 90° and the value of e_N will be less than about 0.2, particularly if the ditch is covered with gravel. Because of the low value for e_N, trajectory heights will be low, with falls rolling rather than bouncing after the first impact.

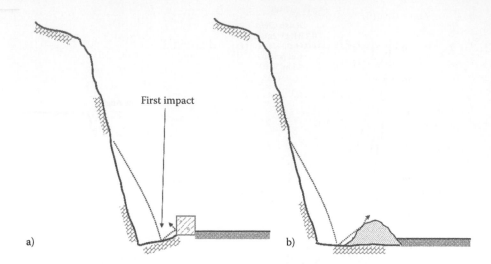

a)   b)

Figure 10.4 Ditch geometry. (a) Catchment area with sloping base and barrier with vertical face. (b) Catchment area with horizontal base and gravel berm with sloping sides.

This behavior of falls after impact in the catchment area demonstrates that a low barrier along the outer side of the ditch will be very effective in containing falls. Furthermore, a barrier with a vertical face will be more effective than one with a sloping face because rocks impacting a vertical face will be deflected back into the catchment area, while they may roll over a sloping gravel berm (Figure 10.4). Another advantage of a barrier with vertical sides is its narrower "footprint" compared to a gravel berm, and the narrower overall width of the catchment area.

The information on the first impact point on Figure 10.3 and the ditch geometry in Figure 10.4 can be used to design ditches for situations, such as railways, where it is permissible to place a barrier in the ditch to contain rolling rock. That is, the minimum required width of the ditch is just beyond the first impact point, with the incorporation of a barrier with a vertical face (Figure 10.4a); this ditch configuration can be much narrower than the catchment areas shown on Figure 10.3. In comparison, the ditch in Figure 10.4(b) with a horizontal base and a gravel berm with sloping sides over which rock falls may roll, requires a greater width.

10.1.3 Gabions

Gabions, which are wire baskets filled with loose, angular rock fragments, are a commonly used means of creating the vertically sided barrier shown in Figure 10.4(a). Gabions are a particularly effective method of containing rock falls because they can absorb impact by the flexibility of the wire mesh and the slight rearrangement of the rock fragments in the baskets.

The optimum rock fill is in the size range of 100 to 150 mm (4 to 6 in.), with no fines so that it drains readily and the gabion does not act as a dam. It is also preferable that the rock fragments are angular, so that high shear resistance between the fragments is developed between the pieces during impact, compared to rounded cobbles. The shear resistance is also enhanced by hand placing and closely packing the rock so that the density and interlocking are maximized.

Advantages of the use of gabions for rock fall control are as follows:

1. Gabions can be constructed from locally available materials that may be supplied at no cost, using rock that will not weather and breakdown with time.
2. Gabions can be constructed on site so that transportation costs are limited to bringing in the wire baskets that are supplied in a flat configuration for ease of handling.
3. Fabrication of the baskets can be directly in the ditch or in a nearby assembly area from which they can be moved to a site with light equipment.
4. If a gabion is damaged by an impact, it can readily be repaired and/or pushed back in place.
5. Plant growth will often occur soon after construction, improving the aesthetic appearance of the barrier with no detrimental effect to its performance.

A few disadvantages of gabions are their labor-intensive construction, and repair can be difficult if damage occurs that breaks open a basket in the middle of a barrier. It is also of interest that the unit weight of a fabricated gabion is about 15 kN · m^{-3} (95 lbf · ft^{-3}) compared to 26 kN · m^{-3} (165 lbf · ft^{-3}) for solid rock. This relatively light weight must be taken into account when examining the impact resistance of gabion barriers, as well as designing retaining structures with gabions.

Figure 10.5 shows examples of barriers constructed with gabions. The barrier in Figure 10.5(a) is constructed with two rows of gabions set down below the level of the highway so that impact resistance is provided by the weight of the fill behind the barrier. The barrier in Figure 10.5(b), constructed on the shoulder of a railroad, was impacted by a boulder with enough energy to displace the upper row of gabions. However, the combined resistance of the deformation of the gabions themselves and the shear force on the base of the structure has been sufficient to absorb the total impact energy and stop the fall with no damage to the railway.

The dimensions of individual gabion baskets are usually 1 m by 1 m (3 ft by 3 ft) in cross-section and they can be stacked to form larger barriers, as required. The energy-absorbing capacity of the barrier is enhanced by placing the individual baskets in a brick-work pattern so that continuous joints are avoided.

Figure 10.5 Examples of rock fall barriers constructed with gabions. (a) two-level wall set below road level and backed with fill; (b) top layer of gabions displaced by rock fall impact.

10.1.4 Concrete Block Barriers

Section 9.3 on the energy-absorbing characteristics of protection structures, and Figure 9.2 demonstrate that rigid, mass concrete walls are not effective rock fall barriers because they shatter rather than absorb energy. However, walls constructed with individual concrete blocks—for example, 1.5 m (5 ft) long and 0.75 m (2.5 ft) square section in Figure 10.6— can be an effective barrier because they have a vertical face to deflect the rocks back into the ditch, and the blocks can move individually to absorb energy. For the barrier shown in Figure 10.6, shear resistance is generated primarily by the weight of the blocks and friction on the base that includes an indented keyway in contact with the granular ground surface. In addition, a steel cable run through the lifting eyes on the top of the blocks both prevents excessive displacement and disperses the impact load over several blocks.

An effective application for concrete blocks such as those shown in Figure 10.6, is to contain falls from loose gravel and boulder slopes that have formed just steeper than the angle of repose (slope angle of about 35° to 37°). Under these conditions, the rocks roll with low trajectories and one or two levels of blocks provide effective protection, particularly if they are set into the valley side of the ditch so that displacement is restrained by fill on the outer side (see Figure 10.5[a]). Although larger walls can be built by stacking blocks using the keyways, the rigidity of the walls increases with their mass resulting in the blocks being more likely to shatter than displace. Further advantages of using concrete blocks for barrier construction are rapid construction, and ease of repair to either reset the blocks in position or replace shattered units.

10.1.5 Impact Energy Capacity of Gabions and Concrete Blocks

The gabion and concrete block barriers described in Section 10.1.3 and 10.1.4, respectively, have limitations as regards their height and energy-absorbing capacity because they will

Figure 10.6 Barrier constructed with individual concrete blocks, with five blocks displaced and one fractured by the impact of the rolling boulder shown in the lower image.

become unstable at large height-to-width ratios; the maximum height may be about 2 m (6.6 ft). The images in Figures 10.5(b) and 10.6 show that significant displacement has been caused by these two rock falls, for which the approximate impact energies can be calculated as follows.

The rock fall in Figure 10.5(b) has dimensions of 2.5 m by 1.5 m by 0.75 m (8 ft by 5 ft by 2.5 ft) for a mass of about 7500 kg (16,550 lb). This rock rolled across a horizontal, vegetated soft soil surface before impacting the gabion at an estimated velocity of 8 to 10 m · s^{-1} (25 to 35 ft · m^{-1}). For these conditions, the possible range of impact energies is 250 to 400 kJ (90 to 150 ft tonf).

The rock fall in Figure 10.6 is approximately cubic shaped with side dimensions of 1.5 m (5 ft), for a mass of about 9,000 kg (20,000 lb). This rock rolled down a 20 m (65 ft) high, angle of repose (37°) slope before landing in the narrow ditch and impacting the concrete block at an estimated velocity of 10 to 12 m · s^{-1} (35 to 40 ft · m^{-1}). For these conditions, the possible range of impact energies is 450 to 550 kJ (165 to 200 ft tonf).

Based on the two rock fall examples shown in Figures 10.5(b) and 10.6, the maximum energy that these two simple barrier types can withstand is about 500 kJ (185 ft tonf). This calculated impact energy represents approximate ultimate limit states energy for these barrier types, or a service limit energy of about 150 kJ (55 ft tonf), since (SEL = UEL/3)

10.2 MSE BARRIERS

High-impact energy-capacity barriers can be constructed using mechanically stabilized earth (MSE), the dimensions of which can be suited to the required energy capacity and anticipated rock fall trajectories. For example, barriers have been constructed with heights up to 15 m (50 ft) and energy capacities of 15,000 to 20,000 kJ (5,500 to 7,400 ft tonf) (Maccaferri, 2012).

The impact energy capacity of MSE barriers is directly related to their dimensions and the method of construction. That is, the required barrier mass is defined by the width and height, and the required resistance to shear displacement within the barrier is achieved by the use of compacted, granular soil, and horizontal plastic or steel-mesh reinforcement. Also, the MSE construction method can readily accommodate irregular foundation surfaces, and a variety of cross-sectional shapes to suit site conditions.

Possible disadvantages of MSE barriers are the space required for the footprint of the structure, together with a catchment area on the mountainside, and the need for a stable foundation to support the dead and live loads of the structure. In addition, access for heavy equipment is necessary for both construction and removal of accumulated rock falls. However, where adequate space is available for barriers, this type of protection is usually economical for both construction and maintenance, especially where all work can be carried out at the level of the highway, for example, in areas readily accessible by equipment.

This section describes features of MSE barrier design and construction.

10.2.1 MSE Barriers—Design Features

Figure 10.7(a) shows an image of an 8 m (26 ft) high MSE barrier designed to contain rock falls with impact energies up to 10,000 kJ (3,700 ft tonf), falling from a slope several hundred meters high; Figure 10.7(b) shows details of the construction method. It is of interest that the barrier in Figure 10.7 was constructed to replace the concrete wall shown in Figure 9.2(b), and Figure 4.3 shows a rock fall impact on the face of the barrier; the approximate impact energy of this rock was 1,200 kJ (450 ft tonf).

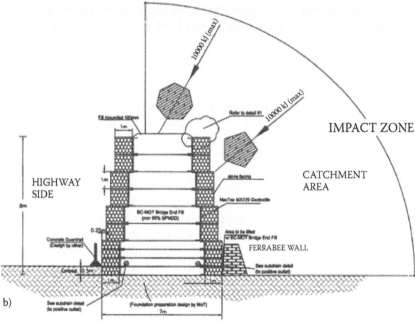

Figure 10.7 Rock fall barrier. (a) 8 m (26 ft) high MSE wall with gabions forming facing surfaces; (b) details of MSE construction, Trans-Canada Highway near Boston Bar, British Columbia. (From British Columbia, MoTI; Maccaferri, 2012.)

Features of typical rock fall barriers using MSE construction methods are discussed below (see Figure 10.8).

- **Steep face angles**—The steep faces (ψ_f), at 84° in the case of the MSE barrier in Figure 10.7, minimize the footprint width of the structure and the volume of construction material required. In addition, the steep mountainside face ($\psi_f > 65°$) redirects rock falls into the catchment area and prevents them from rolling over the barrier. It is not necessary that the two face angles are equal.
- **Foundation**—The foundation is recessed to a depth of about one half the height of a facing element to improve the shear resistance along the base of the structure. Weak

foundation material is removed and replaced with compacted granular fill, if necessary, to ensure that it has adequate bearing capacity to support the weight of the structure and impact loading, without settlement or movement. Another possible foundation condition, where the barrier is located at the crest of a steep slope, is weathering of rock at the crest that undermines the outer edge of the barrier.

- **Design height**—The total height of the structure comprises two components. First, the height, H_1 required to stop rock falls, depending on the expected maximum trajectory height. Second, the upper portion, H_2 of the barrier, above the top impact point, provides a weight that generates shear resistance on horizontal surfaces at the level of the impacts.
- **Crest width**—The minimum crest width C is about 0.8 m (2.6 ft) for the operation of compaction equipment and construction safety. This width is increased as required for increasing the mass and impact energy capacity and is 3 m (10 ft) wide for overall barriers heights greater than 6 m (20 ft).
- **Facing elements**—The functions of facing elements are to retain the fill material forming the body of the barrier, to distribute the impact force over a wide area in order to minimize concentrated pressures in the fill, and to withstand damage from impacts. Common facing materials that fulfill these functions include gabions (steel wire-mesh baskets filled with rock) and geofabric reinforced with steel wire mesh that has better impact resistance and is easier to maintain. Other facing materials have been used successfully as discussed in the following paragraphs.

The gabions on the mountainside face of the MSE barrier shown in Figure 10.7 are protected with a layer of rubber conveyor belting that is more resistant to damage from the high-velocity impacts than the gabions. In Figure 2.1, the 6.6 m (21.5 ft) high MSE

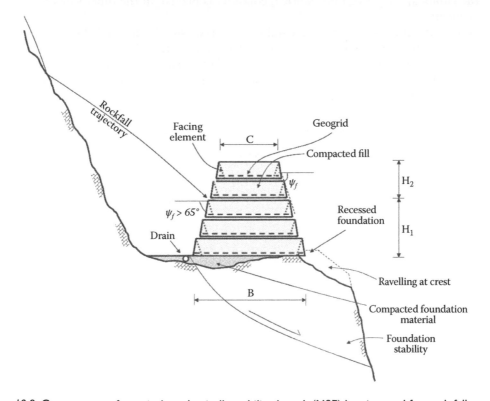

Figure 10.8 Components of a typical mechanically stabilized earth (MSE) barrier used for rock fall control.

wall has precast concrete blocks forming the two faces of the barrier. The mountain-side facing of this MSE barrier has been impacted multiple times by falls with energies up to about 80 kJ (30 ft tonf), and the damage from these impacts has been limited to minor chipping of the concrete (see Section 7.5.1).

Figure 10.9 shows a MSE barrier with the mountainside face protected with bags of sand that can be replaced if damage occurs (Protec Engineering, Japan, 2012). The impact energy capacity of this MSE barrier is up to 5,000 kJ (1850 ft tonf).

- **Reinforcement**—The horizontal reinforcement elements within the fill are usually spaced vertically at about 0.7 to 1 m (2.3 to 3.3 ft) and may comprise either geosynthetic Geogrid, or double-twist steel wire mesh. The reinforcement is connected to the facing elements to form a coherent structure.
- **Fill material**—The material forming the body of the barrier between the facing elements is a free-draining, clean, angular fill with a grain size distribution of about 5 to 50 mm (0.2 to 2 in.). Typically, the fill is placed in maximum 300 mm (12 in.) thick layers and compacted to a Proctor density of 95%. The purposes of compacting the fill are to increase its density and the mass of the barrier, in order to enhance its shear strength and energy-absorbing characteristics. With respect to the energy absorption, a loose fill will be displaced during impact with less absorption of energy, compared to a denser fill.
- **Internal stability of MSE barrier**—Internal stability would be achieved with the appropriate use of horizontal reinforcement and compacted granular fill typical for MSE construction.
- **Fill containment**—For barriers faced with gabions, the fill in the body of the structure will be finer than the rock fill in the gabions. To control migration of fines into the gabions, a layer of filter fabric (geofabric) is placed on the inner side of the facing elements.
- **Global stability**—If necessary, a stability analysis should be carried out to determine the factor of safety against sliding through the foundation. Foundation conditions that may be vulnerable to instability are cut-and-fill construction where the valley side of the MSE barrier is partially founded on fill. In the stability analysis, the MSE barrier would be treated as a surcharge on the bearing surface.

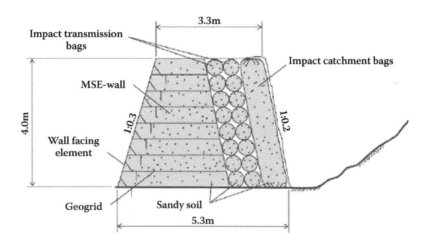

Figure 10.9 MSE barrier with sets of sand bags on the mountain side face that absorb impact energy and can be replaced if damaged. (Courtesy of Protec Engineering, 2012).

- **Aesthetics**—Where the visual appearance of the MSE barrier is of importance, facing elements can be used that enhance the growth of vegetation. This will usually have no detrimental effect on rock fall containment.

10.2.2 MSE Barriers—Design Principles

This section describes the design principles for MSE barriers based on the work of Grimod and Giacchetti (2011). A component of the design depends on proprietary information on the impact energy–deformation behavior of reinforced soil so it is not possible to describe the complete design method. However, the following discussion, together with the design factors listed in Section 10.2.1 above, will show the required approach to MSE barrier design.

The resistance to impact energy of a MSE barrier depends on three factors:

- **Overall mass**—Barrier must have sufficient mass to prevent overall displacement due to impact loading. In comparison, Figures 10.5(b) and 10.6 show that the entire barrier can be displaced by impact forces that are large relative to the barrier mass.
- **Load-deformation behavior**—The barrier material must be able to deform plastically in the region of the impact while the remainder of the structure remains intact. In comparison, Figure 9.2(b) shows how a brittle, unreinforced concrete wall shatters on impact.
- **Internal shear strength**—The concentrated punching effect of the impact must be resisted by the shear strength of the compacted fill and horizontal reinforcement making up the body of the structure.

Studies of MSE barrier behavior under impact loading using finite element analysis and *in situ* behavior, together with calibration from full-scale testing, shows the two primary mechanisms by which the impact energy is absorbed by the structure. First, about 80% to 85% of the energy is absorbed by plastic deformation of the fill material around the impact crater, and second 15% to 20% is absorbed by shear displacement and frictional resistance within the body of the structure. Not more than 1% of the energy is absorbed by elastic movement of the fill (Grimod and Giacchetti, 2011).

The complete design of a MSE barrier involves the study of base sliding and overturning, punching resistance and global/foundation stability as described in the following sections.

10.2.3 Base Sliding and Overturning Stability

The stability of a MSE barrier can be checked for sliding on the base and for overturning using limit equilibrium methods (Wyllie and Mah, 2002), where the factor of safety *FS*, is defined by the ratio of the resisting and displacing forces:

$$FS = \frac{resisting\ force}{displacing\ force} \tag{10.1}$$

where the resisting force is either the shear strength on the base, or the returning moment.

In the case of barriers, the displacing force is the impacting force of the rock fall defined by its magnitude and angle, and the impact height above the base.

An approximate method for defining the impact force is as follows. From tests to determine the impact capacity of rock shed roofs protected by a sand cushion, the energy of a block of rock impacting the sand cushion can be expressed as a static force, F (Yoshida et al., 2007). These tests involve dropping, from a height H, blocks of rock with mass m (kg) and diameter D on to concrete roof slabs protected with a layer of sand with thickness T. Accelerometers in the rock and pressure meters on the concrete roof measured the force F transmitted through the sand in to the concrete slab (see Chapter 11 discussing the design of rock sheds). The magnitude of the transmitted force F is equal to

$$F = 0.02(m \cdot g)^{0.67} \lambda^{0.4} H^{0.6} \left(\frac{T}{D}\right)^{-0.58} \tag{10.2}$$

where λ is the Lamé parameter of the cushioning material, which has value of approximately $3,000 \, \text{kN} \cdot \text{m}^{-2}$ for compacted fill. For fill with deformation modulus E and Poisson's ratio v, the Lamé parameter is given by:

$$\lambda = \frac{E \cdot v}{(1+v)(1-2v)} \tag{10.3}$$

The ratio (T/D) will be at least five for a stable barrier; that is, with the horizontal width at least five times the block diameter, so the value of last term in Equation (10.2) can be approximated to $(T/D)^{-0.58} = 0.4$. Equation (10.2) relates the impact force to the free-fall height H, whereas for a MSE barrier the impact velocity, v will be determined from the rock fall trajectories. For free fall velocity, $v = \sqrt{2 \cdot g \cdot H}$, or $H = (v^2/2g)$. Using this expression for H, the value for $H^{0.6} = (v^{1.2}/(2 \cdot g)^{-0.6}) = (0.17 \cdot v^{1.2})$, and the ratio $(T/D = 5)$, Equation (10.2) can be expressed as

$$F \approx 0.14(m \cdot g)^{0.67} \lambda^{0.4} v^{1.2} \tag{10.4}$$

Figure 10.10 shows a MSE barrier with a weight $(m \cdot g)$, face angle ψ_f and base width B impacted by a force F at angle ψ_i and a height above the base H_1. If the friction angle on the base of the MSE barrier is φ, then by limit equilibrium methods, the factor of safety FS against sliding of the MSE barrier on its base is given by

$$FS = \frac{(m \cdot g + F \cdot \sin \psi_i)\tan\phi}{F \cdot \cos \psi_i} \tag{10.5}$$

The overturning stability can be examined by taking moments about O, at the valley side base of the barrier, with the structure being stable if

$$\left[m \cdot g \frac{B}{2} + F \cdot \sin \psi_i \left(B - \frac{H_1}{\tan\psi_f} \right) \right] \geq [F \cdot \cos \psi_i \cdot H_i] \tag{10.6}$$

10.2.4 Punching Stability

Figure 10.11 shows the typical behavior of a MSE barrier under impact loading based on the field testing and theoretical studies. That is, the impact creates a crater on the mountainside with depth δ_m and shear displacement occurs on two horizontal surfaces within the body of the barrier. If the impact force is sufficiently large, the shear movement extends over the full width of the structure and displacement δ_v occurs on the valley side, with $(\delta_m > \delta_v)$ by the amount of the plastic deformation of the fill.

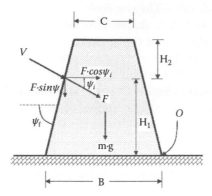

Figure 10.10 Parameters used in calculating factor of safety against base sliding and overturning stability of barrier.

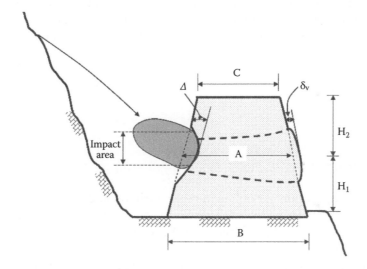

Figure 10.11 Impacted barrier showing crater in mountainside face, depth Δ and displacement of valley side face, δ_v. (From Grimod, A. and Giacchetti, G. (2011). Protection from high energy impacts using reinforced earth embankments: Design and experience. *Proc. Second World Landslide Forum*, Rome, October.)

As a comparison, Figure 4.3 shows a block of rock impacted into the face of the barrier shown in Figure 10.7; this impact did not cause displacement on the valley side.

The behavior of a MSE barrier under punching impact as shown in Figure 10.11 can be used to design these structures to withstand a specified impact force. The sequence of calculations, as described below, shows the general design procedure.

a. **Face angles, ψ_f**—Selected values for face angles are usually in the range of 65° to 80°. Steep faces give the best combination of small footprint and construction volume, adequate stability, and the ability to prevent rotating rock falls from rolling over the structure.

b. **Impact height, H_1**—Rock fall analysis and trajectory calculations will show the maximum likely impact height on the barrier, which is designated H_1. An additional height H_2 is required, for a total height of ($H_1 + H_2$), to provide a normal force on the area of impact and generate shear resistance to the punching load.

c. **Mass and shape of rock fall**—The site geology will show the likely maximum dimensions of the rock falls and their shape. This information can be used to calculate the mass of the block and the area of the impact footprint on the face of the barrier.

d. **Impact velocity**—The trajectory analysis will provide information on the impact velocity, which is related to the fall height, the slope angle, and the effective friction of the slope surface as discussed in Section 3.2.2 and shown in Equation (3.13).

e. **Impact energy, E_i**—The mass, shape, and velocity values can be used to calculate the impact kinetic energy (KE = ½ $m \cdot V^2$) and the rotational energy (RE = ½ $I \cdot \omega^2$).

f. **Mountain side displacement**—Displacement on the mountain side Δ is the sum of the penetration of the block into the barrier δ_m, and the valley side displacement due to shear sliding of the impacted layers, δ_v and is given by:

$$\Delta = \delta_m + \delta_v \tag{10.7}$$

g. **Impact crater depth, δ_m**—The volume of the crater is related to the impact energy E_i, the footprint area of the impacting block on the face, and its shape, and the properties of the reinforced fill forming the barrier. The depth of the crater on the mountainside δ_m is related to the plastic deformation of the fill resulting from the impact.

$$E_i = \frac{volume\, of\, crater}{k \cdot p_c}$$

or

$$\delta_m = \frac{\chi \cdot E_i \cdot p_c}{(footprint\, area\, of\, impact\, on\, face)} \eta \tag{10.8}$$

where the design parameters are defined by (Grimod and Giacchetti, 2011):

- k is the correlation coefficient between the crater volume and the energy developed to form the crater, which depends on the reinforcement material, and the type and compaction of the soil;
- $p_c \approx 0.85$, the approximate portion of the impact energy dissipated by the formation of the crater;
- χ is a proprietary function, determined from finite element analysis, related to the type of reinforcement;
- η is a function of the block shape: $\eta = 1.0$ for cube, $\eta = 1.2$ for sphere.

h. **Valley side displacement, δ_v**—The valley side displacement is related to the impact energy E_i and the friction force generated in the reinforced fill due to sliding:

$$\delta_v = \frac{E_i \cdot s_c}{friction\, force} \tag{10.9}$$

where $s_c \approx 0.15$, is the energy dissipated due to friction within the reinforced fill. The friction force is generated between the compacted soil and the reinforcement.

Once the crater depth has been established from Equation (10.8), and the valley side displacement from Equation (10.9), the total mountain side displacement Δ, is found from Equation (10.7). An approximate relationship between these three displacement parameters has been determined from numerical analyses of impact and displacement, as follows:

$$\frac{\delta_v}{\Delta} \approx 0.6 \text{ to } 0.8 \tag{10.10}$$

i. **Width of barrier, A at impact point**—The performance of actual barriers and the numerical analysis, shows that the width of the barrier A, at the level of impact is related to the value of Δ as follows:

$$A \not< 5 \Delta \tag{10.11}$$

As a design guideline, the generally accepted values for the maximum displacements are:

$$\Delta \not> 0.5 \text{ to } 0.7 \text{ m (1.6 to 2.3 ft); } \delta_v \not> 0.3 \text{ to } 0.4 \text{ m (1 to 1.3 ft).}$$

Combining these limiting values for δ_m and Equation (10.11) shows that the minimum width A would be about 3.5 m (11.5 ft). The guideline for the width A given by Equation (10.11) can be compared with actual design values. For example, Figure 10.7b) shows the cross-section of an 8 m (26 ft) high barrier where the width at the impact level is about 6 m (20 ft). Similar dimensions are shown for the embankment in Figure 10.9.

An adequate barrier width at the impact point ensures that the extent of the displaced fill materials is restricted to the impact area, and that the overall barrier remains intact.

j. **Top width barrier, C**—The required top width of the barrier is related to the depth of the impact crater δ_m, to ensure that sufficient mass exists in the upper part of the barrier to resist punching failure. The minimum value of C is 0.8 m (2.6 ft), which is the required operating width for compacting equipment and worker safety, but the value of C is also related to the crater depth Δ as follows:

$$C \geq 2 \Delta \tag{10.12}$$

where Δ is found from Equation (10.7).

k. **Barrier height, $(H_1 + H_2)$ and base width, B**—Once the widths at the top and the impact point have been established, together with the maximum impact height H_1, the overall height and base width can be determined from the face angles and the structure geometry.

10.2.5 Global Stability

Barriers may be located on benches cut into slopes, or on low-strength foundation materials where the additional load from the structure may result in shear sliding on surfaces under the barrier. Figure 10.8 shows a typical circular sliding surface through the barrier foundation. Possible remedial measures include removal of low-strength foundation materials and replacement with compacted fill.

Global and internal stability can be evaluated using common circular failure type computer programs with the slope model incorporating the barrier geometry with a density consistent with the degree of compaction of the fill. A free stability program for the design of reinforced soil structures is also available from Maccaferri.

10.2.6 Repairs to Face Elements

The suggested limits to the crater depth and valley face displacement of about 0.7 and 0.4 m (2.3 and 1.3 ft), respectively (see Section 10.2.4, item [h] above), are primarily related to stability of the structure. However, these limits also facilitate repair of the damaged area at the impact point. A method of repairing impact damage is to cut away the damaged mesh, fill the crater with sandbags, and then attach a new piece of mesh, with wire clips, to the surrounding intact mesh. Damaged horizontal reinforcement strips should also be repaired and attached to the facing elements.

10.3 SLIDE DETECTOR FENCES

Slide detector fences comprise a series of posts at the base of the rock face that support electrical wires strung horizontally along the alignment (Figure 10.12). For very steep slopes, the tops of the posts may also incorporate cross beams supporting wires that will detect rocks falling vertically from the face. When a rock fall breaks one of the wires, a signal located beyond the stopping distance of the train or vehicles is activated to halt traffic. It is usual practice on railways for trains to then proceed at slow speed so that they can either stop at the location where the wire is broken or proceed if the line is clear.

Slide detector fences are commonly used on freight railways where the traffic is controlled by signal systems, and their speed allows them to stop within distances of about 1 to 2 km

Figure 10.12 Slide detector fence for a railway comprising timber posts supporting wires that, if broken by a rock fall, activate signals to stop trains.

(0.6 to 1.2 miles). Detectors may also be used on very low-traffic-volume roads to warn drivers of events such as rock slides and snow avalanches.

While detectors are easy to construct and maintain, they have the disadvantage that they can be activated by small rock falls and snow or ice slides that are not large enough to disrupt traffic. Also, traffic is not protected after it has passed the signal that, for a railway, would be located as much as 2 km (1.2 miles) from the hazard area to allow for their long stopping distance. It is also necessary to have a source of power to operate the detector fence.

As a replacement for detector fences, tests have been conducted of cables buried beside the track to detect vibrations induced by rock falls. The basic requirement for this system is the ability to distinguish the vibration signature of rock falls from the many other sources of vibrations produced on an operating railway (RGHRG, 2012). Also, it is necessary that the cables be buried below the depth of possible disturbance from track maintenance equipment.

10.4 WIRE MESH—DRAPED AND PINNED

The barriers and embankments discussed in Sections 10.1 and 10.2 require that space be available at the base of the slope for both the footprint width of the barrier itself, and the catchment area behind the barrier. Where space at the base of the space is not adequate for ditch excavation or embankment construction, suitable protection measures may be draped wire mesh or a fence. This section describes the use of steel wire mesh for rock fall protection and discusses the appropriate applications for draped mesh that is suspended from the crest of the slope, and pinned mesh that uses pattern rock bolts to anchor the mesh to the face. The applications of these alternative designs are as follows:

Draped mesh—the mesh is supported only along the crest of the rock slope, with no anchors on the face, and is open at the lower end so rock falls are directed by the mesh into a catchment area beside the road. Draped mesh is suitable for locations where space for a catchment area is available, and the rock face is generally stable except for occasional rock falls. If bolts are required in localized areas of instability, they should be installed before the mesh is hung to avoid having to remove mesh if bolting is needed later.

Pinned mesh—The mesh is pinned to the face with a pattern of rock bolts so that the bolts and mesh together support loose rock on the face. The spacing and diameter of the bolts and the strength of the mesh is suited to the weight of the loose rock to be supported.

These two mesh designs are discussed in Sections 10.4.1 and 10.4.2 below.

10.4.1 Draped Mesh

A simple method of containing rock falls is to hang steel wire mesh on the rockface from a support system located along the crest (Figure 10.13). Figure 10.13(b) shows details of the support, with the top of the mesh raised about 1 m (3.3 ft) above the ground so that rocks falling/rolling from the slope above the top of the mesh will be contained. The benefit of draped mesh is that its combination of weight and flexibility allows it to hang close to the face to restrain falling rocks so that their velocity, and impact on the mesh, is limited. This allows a lightweight mesh to provide effective protection on many slopes. However, the maximum size of falls that can be contained by unreinforced draped mesh, without damage, is about 0.5 to 1 cu. m (0.7 to 1.3 cu. yd).

As shown in Figure 10.13(a), the lower edge of the mesh is open and above the level of the ditch which allows rocks to fall down the face behind the mesh. It has been found that if the mesh extends into the ditch, rock falls tend to collect on the lower edge of the mesh that can be torn by the weight of the accumulated falls.

Figure 10.13 Draped wire mesh. (a) Mesh suspended from crest of excavated rock face; (b) detail of support system at crest of cut (Sea to Sky Highway, British Columbia).

Types of mesh that are appropriate for this application include various chain-link meshes with wire diameters of about 2 to 4 mm (0.08 to 0.16 in.). Suitable meshes that will not unravel if a wire is cut have double twists at each connection, or high-strength wires with resistance to bending. It is also possible to increase the overall strength of the mesh by incorporating steel-wire cables woven into the mesh at a spacing of about 4 to 20 times the spacing of the mesh wires.

High-strength mesh and heavy-duty support are often required where the dimensions of the potential rock falls are substantial, and for installations on high, near vertical slopes where the weight of the mesh itself, hanging freely down the face, is a significant load. On flatter slopes where the mesh lies partially on the face, a lower strength system may be adequate.

Mesh installations should always consider the required corrosion resistance of the steel wire, depending on such factors as the design life, proximity to the ocean or sources of acidic industrial air emissions, and rainfall intensity. Most commercial mesh products are galvanized and the product properties should be suited to site conditions.

Mesh is usually supplied in rolls, and some types of mesh retain a curved shape when unrolled at the site. When the curved mesh is rolled down the face, it tends to hang at some distance from the face, resulting in limited confinement of falling rock. In order for the suspended mesh to hang close to the face, it is necessary to reroll the mesh in the opposite direction to that in the supplied rolls to take out the curve (termed "back rolling").

10.4.2 Mesh Pinned to Face with Pattern Bolts

Mesh pinned to the face with pattern rock bolts is suitable for locations where no catchment area is available at the base of the slope, and it is necessary to contain rock falls where they loosen on the face. The pinned mesh provides support of the rock face by the combination of the rock bolts installed in potentially unstable blocks of rock, and the resisting force developed by the mesh tightened against the face by the bolts. Tension in the mesh is achieved by locating the rock bolts in hollows in the rock face and then placing the mesh under the plates and torquing the nuts and plates tightly against the face.

A particular advantage of slope support with pinned mesh is its generally good performance during earthquake ground motions because of the flexibility of the system. That is, the mesh will provide support during the earthquake and rock falls that do occur will be contained by the mesh.

Pinned mesh can support either individual blocks of rock that may slide from the face or accumulations of smaller rock fragments. Figure 10.14 shows an example of pinned mesh installed at the crest of a steep face in weathered basalt. The rock is strong but closely jointed resulting in ongoing ravelling-type instability of rock fragments that were a hazard to pedestrians below the slope. The mesh was effective in containing these falls, and the bolts and mesh together had sufficient strength to stabilize the face.

Figure 10.14 Example of mesh pinned to face to contain strong, closely fractured rock (Stanley Park, Vancouver).

The design of a pinned mesh installation for blocky rock involves the use of limit equilibrium methods in which the driving force produced by sliding block of rock is resisted by the combined shear strength on the sliding surface and tension in the mesh draped over the block. The magnitude of the mesh resisting force is related to the tension in the mesh and load-deformation characteristics, with the resisting force increasing with the stiffness of the mesh. That is, a highly flexible mesh will deform as the blocks slides and will provide little resistance to prevent the block from being dislodged from the slope. However, a stiff mesh will provide resistance as soon as the block moves.

Information required for the design of a pinned mesh installation for blocky rock is the characteristics of discontinuities that could form blocks on the face, including their orientation, spacing and persistence, and the shear strength of potential sliding surfaces. This information, which would be obtained by mapping the rock face or nearby outcrops, will determine the likely shape and dimensions of the blocks, and the dip of the sliding plane. For closely jointed, ravelling rock, the required design information is the depth of the loosened rock behind the face or the expected volume of rock that will accumulate behind the mesh.

Other design information for pinned mesh is the tensile strength of the wire-mesh and its load-deformation characteristics. These properties of various wire mesh products have been determined from laboratory testing, as well as full-scale testing of pinned mesh using a hydraulic ram mounted on the rock face to simulate a block of rock sliding from the face (Giacchetti et al., 2011). These tests show that the rock bolt spacing should be about 1.0 to 1.5 m (3.3 to 5 ft), and never greater than 3.0 m (10 ft), in order to distribute the rock load reasonably uniformly in the mesh and bolts.

The principles of pinned mesh design using limit equilibrium methods are as follows. Figure 10.15 shows a potentially unstable block of rock, located between two rock bolts, that can slide from the face; the weight of the block is W and the dip of the face is ψ_f. The sliding plane with area A and dipping at angle ψ_p out of the face has shear strength properties of cohesion c and friction angle φ. Mesh is pinned to the face by pattern rock bolts and is tensioned to force T by tightening the nuts on the bolts and by movement of the block. The

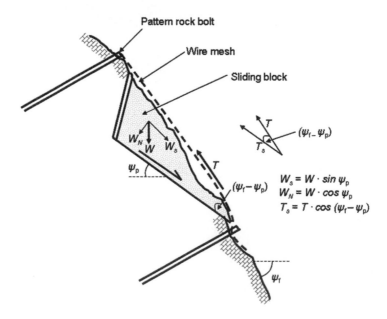

Figure 10.15 Potentially unstable block of rock held in place with tensioned wire mesh pinned to face with pattern rock bolts.

component of the mesh tension acting up the sliding surface, which is a resisting force, is given by $T_s = T \cdot \cos(\psi_f - \psi_p)$. The mesh tension will not generate a normal component of T acting out of the slope because the mesh is not attached to the rock face. It is assumed that no water pressures exist close to the slope face. The factor of safety, FS, of the block against sliding is

$$FS = \frac{resisting\,force}{sliding\,force} \tag{10.13}$$

$$= \frac{c \cdot A + (W \cdot \cos\psi_p)\tan\varphi}{W \cdot \sin\psi_p - T \cdot \cos(\psi_f - \psi_p)}$$

Equation (10.13) gives the tension in the mesh to hold a block of rock in place with a required factor of safety.

Computer programs available for designing pinned mesh systems are available from Maccaferri (BIOS and Mac.Ro1) (Maccaferri, S.p.A., 2012) and Geobrugg (Ruvolum) (Geobrugg, 2012). These programs can examine a variety of loading conditions from blocky rock to granular material, and the output from the analysis is the required tensile strength and spacing of the rock bolts and the strength of the mesh.

The designs for pinned mesh only examine the stability of the rock bolt-mesh-surficial rock system and do not determine the required length of the bolts to prevent larger-scale instability. Figure 10.16 shows a pinned mesh installation where slope instability extending beyond the depth of the rock bolts has occurred resulting in the destruction of the mesh. In addition, localized failure of the mesh is occurring where the weight of the accumulated rock fragments has exceeded the strength of the mesh and steel strapping.

Figure 10.16 Mesh pinned to face with pattern rock bolts and steel straps. Fragments of weathered rocks are accumulating behind the mesh and strapping, and a slope failure below the depth of rock bolts can be seen at the right side of the image.

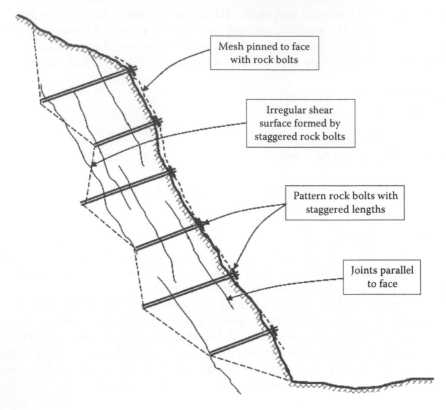

Mesh pinned to face
with rock bolts

Irregular shear
surface formed by
staggered rock bolts

Pattern rock bolts with
staggered lengths

Joints parallel
to face

Figure 10.17 Pinned mesh with staggered bolt lengths to create an irregular shear surface beyond the ends of the bolts.

The image in Figure 10.16 shows that the rock slide took place on a set of discontinuities oriented approximately parallel to the excavated face. To ensure the stability of both the pinned mesh system and the overall slope, a stability analysis, independent of the mesh design, is necessary to determine the depth of potential sliding surfaces and required length of the bolts.

It has been found that, if a pattern of rock bolts with all the same length are installed, a plane of weakness on which movement can occur may be created just below the ends of the bolts. However, the installation of bolts with staggered lengths creates an irregular surface that does not lie on a single discontinuity (Figure 10.17). For the irregular surface created by the staggered bolt lengths, the shear strength is improved because shearing on this surface has to occur primarily through intact rock rather than along discontinuities.

10.5 NETS AND FENCES

A wide variety of steel-mesh fences are commercially available with rated impact energy capacities of between 100 kJ (40 ft tonf) and 8,000 kJ (3,000 ft tonf). A common characteristic of all these products is the flexibility of the overall structure such that the impact energy is absorbed by deflection and strain of the fence components according to the design principles discussed in Chapter 9.

The basic design parameters of all fences are the design impact energy based on the rock fall mass and velocity, and the height based on the rock fall trajectories. These values would be determined using trajectory calculations as discussed in Chapter 3, or by rock fall modeling as discussed in Chapter 7. Consideration should also be given to the fence location where the energy and trajectory height are minimized. That is, ideal locations are in gullies where rock falls are concentrated (see Figure 3.12) or on benches where the energy is minimized by the fall impacting the ground just before it impacts the fence (see Figure 9.1).

A design consideration related to the flexibility of the fence is its location in relation to the facility that it is protecting. In general, the fence should be low on the slope to facilitate construction and maintenance. However, the distance between the fence and the facility must allow clearance for deflection of the net on impact so that vehicles, for example, do not run into rocks contained in a deflected net. Fence manufacturers provide deflection information based on the results of full-scale testing.

10.5.1 Fence Components

A typical fence installation is shown in Figure 10.18 and its common features are discussed in the following paragraphs.

 a. **Posts**—The posts supporting the net are steel beams, or sometimes pipes, spaced at about 8 to 12 m (25 to 40 ft) and with heights up to about 8 m (25 ft). The posts incorporate brackets to which the support cables for the net are attached. The strength of the posts is adequate to support the net components and to sustain the impact loading of rock falls into the system. If possible, the posts should be located outside rock fall gullies so that most impacts are on the flexible net rather than the rigid posts.

   The layout of the posts influences the construction and performance of the fence. It is preferable that the fence is approximately linear in plan because sharp changes in

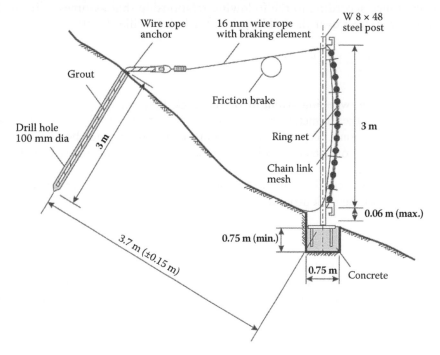

Figure 10.18 Typical rock fall fence with rigid base in soil, up-slope guy wires and friction brakes. (Courtesy of Geobrugg Inc.)

direction require extra guy wires and anchors to support the post where the change in alignment occurs. It is also preferable that the posts are the same length and that no sudden change in elevation occurs between two posts. For sites with uneven topography, gaps between the bottom of the net and ground may occur, requiring the use of custom-shaped nets and different post lengths. Alternatively, a series of short, straight fences could be constructed, each suited to the topography and with uniform net panels and post lengths.

As discussed in Section 9.4, in the ideal fence design, all the components—net, brakes, and hinges—have the same stiffness such that energy is absorbed equally by each component of the structure.

b. **Foundations**—The type of foundation will depend on the bearing materials, either rock or soil, and the system design loads.

For *rock foundations*, the usual procedure is to install a group of fully grouted rock bolts drilled into the rock to which the base of the support post is bolted. The design procedure for the rock bolts would be to check the strength of the bolts in shear and tension, the rock-grout pull-out resistance, and the stability of the cone of rock developed by the group of rock bolts (Wyllie, 1999). It is preferable that the bolts be tensioned equally and locked-off against the base plate so that the foundation is loaded in compression and shear resistance generated between the base plate and the rock surface minimizes the shear forces in the bolts.

The strength and number of rock bolts is related to both impact energy and whether the base is hinged or rigid. Flexible hinged posts help to absorb impact energies, whereas rigid posts transfer impact forces to the foundations resulting in higher foundation loads. It is usual that bolt diameters of 25 mm to 32 mm (0.9 to 1.3 in.) will provide adequate resistance.

For cement-grouted bolts, the length of the bolts L_b will be determined by the pull-out resistance according to the following relationship that assumes a linear distribution of shear stress over the full length of the bolt (Wyllie, 1999):

$$L_b = \frac{S_w}{\pi \cdot d_h \cdot \tau_{all}} \tag{10.14}$$

where S_w is the working strength of the rock bolts (usually 50% of the ultimate strength), d_h is the diameter of the drill hole, and τ_{all} is the allowable shear stress of the rock-grout bond. Typically, the strength of the rock-grout bond is less than the strength of the grout-steel bond so the rock-grout bond strength will be the relevant design parameter.

The working strength of the rock-grout bond is related to the uniaxial compressive strength $\sigma_{u(r)}$ of the rock around the periphery of the drill hole and is approximately:

$$\tau_{all} = \frac{\sigma_{u(r)}}{30} \tag{10.15}$$

For example, for a 28-mm-diameter bolt with $S_w = 220$ kN, installed in a 80-mm-diameter drill hole in strong limestone with $\sigma_{u(r)} = 55$ MPa, the allowable bond stress is $\tau_{all} = 55/30 \approx 2$ MPa. From Equation (10.14), the required bolt length $L_b = 220/(\pi \cdot 0.08 \cdot 2000) = 0.4$ m.

While a bolt length of 0.4 m (1.3 ft) is adequate in strong rock to develop the full working strength of the bolt, it would be usual to install 2 m (6.6 ft) long bolts to ensure that they are bonded below any loose or weathered surficial rock. No benefit is usually achieved by installing bolts longer than about 2 m (6.6 ft) because most of the load in a fully grouted bolt is concentrated in the upper about half meter approximately, and no significant load is carried below this depth.

Another design issue for a rock foundation is the orientation of discontinuities that could form potentially unstable wedges of rock that may slide from the foundation. This condition could occur if the foundation is at the crest of a steep slope where the joints dip out of the face. For these circumstances, the bolts should be oriented to cross the joints on which sliding could occur.

Overall, the weakest point in the foundation is usually the base of the posts and the heads of the bolts; it is unlikely that fully grouted bolts will pull out of the rock or movement of the reinforced foundation rock occur.

For *soil foundations*, two possible foundation types are bolts, using either cased holes or self-drilling bolts, or a mass concrete block (Figure 10.16). Self-drilling bolts are hollow-core drill steels with a sacrificial bit that are drilled to the required depth and then left in place. Cement grout is circulated through the hollow core and returned at the surface to grout the bolt in place. Self-drilling bolts are suitable for sites with granular but low permeability soils where the grout will be retained around the bolt. The allowable bond stress for these ground conditions is about 0.05 to 0.1 MPa.

If drilled bolts are not suitable for the site conditions, a mass concrete block could be poured in an excavation at each foundation location. Stability of the foundation would be enhanced by casting a truncated cone and compacting the soil around the concrete to enhance pull-out resistance.

c. **Hinges**—The bases of the posts can either be rigid as shown in Figure 10.18, or hinged as shown in Figure 9.5. A rigid base would be required where it is not possible to use upslope guys to support the posts.

Rigid bases that must sustain impact load on the posts need to be more robust than flexible, hinged bases. As the energy capacity of fence increases, the posts and rigid foundations may become so large that a hinged base is preferred. The advantage of a hinged base is that no damage to the foundation occurs if the post is deflected; repairs are limited to resetting the posts and replacing the brakes on the guy wires (see item g below).

d. **Net**—The type of net used in fence construction depends on the design impact energy, and many proprietary and fully tested systems are available. Where the primary structural net on a rock fall fence has large openings, a secondary chain-link wire mesh is used on the mountainside of the net to stop small rocks that may pass through the larger openings, and to protect the primary net.

It has been found that nets fabricated from high-strength wire are highly resilient to impact loads. For example, at the Mount Stephen rock fall site described in Section 2.1.1 and Figure 2.1, the seven-wire Ringnet was impacted multiple times by falls with velocities up to about 45 m · s^{-1} (150 ft · s^{-1}). These impacts destroyed the chain-link wire mesh on the mountain face of the net, but damage to the rings has been limited to occasional plastic deformation; no rings have been broken.

An issue for design and construction of rock fall fences is the dead weight of the net because this may influences, for example, the capacity of cranes or helicopters to lift the net into place. Examples of net weights are

- Double-twist chain-link mesh—approx. 1.80 kg · m^{-2}
- 3 mm (0.12 in.) wire diameter Tecco mesh—1.65 kg · m^{-2}
- HEA panel barrier—2.30 kg · m^{-2}
- Omega net—approx. 2.7 kg · m^{-2}
- Ringnets with seven-wire loops per ring—5.20 kg · m^{-2}

e. **Net support cable and attachments**—A strong but flexible system is required to attach the net to the support posts. One attachment method is a support cable that attaches to the tops and bases of the posts and runs between the posts forming top and bottom support cables. Depending on the energy capacity, the support ropes can incorporate brakes to help absorb impact energy. The net can be attached to the support cable loop using a "lacing wire" threaded through openings in the mesh. Alternatively, shackles can be used to both attach the net to the support cable and to join net panels.

f. **Guy wires**—Guy wires may be needed upslope of the fence to support the posts if the bases are hinged, and at the ends of the fence to provide support in the longitudinal direction. Many guy wires incorporate brakes, which are described in item (g) below, that absorb energy when the posts deflect and are an important component of the overall flexibility and energy dissipation of the fence system. For efficient functioning of the guy wires, they should be installed close to the line of the applied load.

The steel cables used for the guy wires are attached to the posts at one end and to anchors grouted into the rock or soil slope at the other. The attachment details of the cables as defined by the cable manufacturer are important for making sure that the maximum strength of the cable is operative at the connection points. That is, all loops in the cable should contain a "thimble" that prevents very low-radius, high-stress bends in the cable, and the correct number, positioning, and torquing of the cable clamps should be used (see Figure 10.19[b]). With respect to the positioning of cable clamps, the rule "Never saddle a dead horse" must be applied (i.e., the cable clamp saddle must be positioned on the live end of the cable). For a cable installed with the correct attachment procedures, the strength of the cable is down-rated by about 20% to account for the stresses induced in the cable by the cable clamps and other connection hardware.

g. **Brakes on guy wires and support cables**—Fence manufacturers have developed proprietary brakes to incorporate into guy wires and support cables that contribute to the energy absorption (Figures 10.19[a] and [b]). Types of brakes include loop brakes (Geobrugg), coil brakes (Trumer Schutzbauten), and compression tubes (Maccaferri). When the impact energy exceeds the initiation energy, the brakes are activated and plastic deformation occurs. After rock fall impacts, it may be necessary to replace the brakes, and in active rock fall areas a supply of fence components may be kept available to facilitate maintenance.

h. **Grouted anchors**—The anchors that secure guy cables to the slope should have the same strength as the guys, down-rated as appropriate, with the embedment length determined by Equations (10.14) and (10.15).

The guy anchors often need to be installed at a steep angle so that they are fully embedded in the rock and extend below the depth of weathering. In these circumstances, the anchor and the guy may not be in alignment, resulting in bending of a rigid bolt. An alternative to rigid bolts is to use a steel cable loop anchor comprising a loop of steel cable that can readily accommodate bending. The loops are equipped with a thimble at the upper end and swaged clips to hold the wires together. It has been found that a 3 m (10 ft) long loop of 18 mm (0.7 in.) diameter steel cable, grouted with cement into a 75 mm (3 in.) diameter drill hole is an effective anchorage (see Figure 10.19[b]).

i. **Factors of safety in net design**—The design of rock fall fences will usually require that a number of assumptions be made with respect to such parameters as the design energy, site topography, and analysis method (Giacchetti and Zotti, 2012). In order to quantify this uncertainty, partial factors of safety γ can be incorporated into each component of fence design as follows (ETAG 027):

Figure 10.19 Details of nets, brakes, cable loop anchors, and cable connection systems; (a) ringnet with loop brake (Geobrugg); (b) coil brake and cable attachment details (Trumer).

- Design energy level: SEL approach, γ_E = 1.0, UEL approach, γ_E = 1.2
- Risk to human life: low risk, γ_R = 1.0, high risk, γ_R = 1.2
- Reliability of simulation software: simulation based on back analysis, γ_{Tr} = 1.02; simulation based on bibliography, γ_{Tr} = 1.1
- Quality of topographic survey: high quality, γ_{DP} = 1.02; low quality γ_{DP} = 1.1
- Barrier deformation: SEL approach, γ_d = 1.0; UEL approach, γ_d = 1.3; UEL approach or fence has less than three spans, γ_d = 1.5

SEL: service energy limit; UEL: ultimate energy limit (see Section 8.2).

10.5.2 Attenuators and Hanging Nets

The type of fence shown in Figure 10.18 is designed to stop and contain rock falls and will need to be cleaned if rocks accumulate behind the net. Furthermore, as discussed in Section 9.2 and shown in Figure 9.3(a), the rock fall impact at right angles to the net results in the total impact energy being absorbed by the fence expressed as zero efficiency in Equation (9.1).

In order to produce a structure that is both self-cleaning and only absorbs a portion of the impact energy, attenuator-type structures have been developed as shown in Figures 9.3(b) and 9.4. The concept of attenuation structures is that rock falls impact the net at a shallow angle so that they are redirected by the net, with the velocity being reduced but not diminished to zero. The efficiency (efficiency = final energy/impact energy × 100%) of an attenuator can be as high as 50%.

The hanging net type of attenuator shown in the diagram and image in Figure 10.20 has been used by the author since the mid-1990s to contain falls on steep slopes where the

Figure 10.20 Typical attenuator hanging net that redirects rock falls into the ditch beside the railway track.

available catchment area width is limited, that is, less than a meter. The net consists of a pair of posts either side of a gully, or a series of posts along the slope, that are supported with upslope and longitudinal guy wires, each equipped with a braking element(s). In addition, vertical cables, which are hung from each post and anchored to the rock face at ditch level, are tied to the net panels with shackles or lacing cable. These vertical cables ("side restraint cables") prevent the net panels from sliding inward ("necking") under their own weight, and when the net is impacted by a rock fall.

The tops of the posts are at the same level so the support cable is horizontal and the net does not slide down the cable. Also, the posts are positioned so that the net hangs vertically along the line of the catchment area. With this arrangement of the posts and net, rocks do not accumulate on the slope but fall into the catchment area where they can be readily removed.

The types of attenuator nets shown in Figure 10.20 have been impacted thousands of times by rock falls and have sustained impacts up to possibly up to 1,000 kJ (370 ft tonf) without damage, although controlled testing has not been carried out to determine the maximum design energy.

10.5.3 Debris Flow Barriers

Debris flows are highly fluid mixtures of water, solid particles, and organic matter. This mixture has a consistency of wet concrete and consists of water (possibly 20% to 50%), and solid material ranging from clay and silt sizes up to boulders several meters in diameter. The organic matter can include bark mulch as well as large trees swept from the sides of the channel. Debris flows usually occur during periods of intense rainfall or rapid snowmelt, and a possible triggering event can be collapse of a temporary dam formed by a slope failure or a logjam that releases a surge of water and solid material. Where such flows originate in streams with gradients steeper than about 20° to 30°, they move at velocities of approximately 3 to 5 m · s⁻¹ (10 to 15 ft · s⁻¹), with pulses as great as 30 m · s⁻¹ (100 ft · s⁻¹). At this speed, material is scoured from the base and sides of the channel so the volume of the flow increases as it descends. This combination of high density and high velocity can cause devastation to any structure in its path (Skermer, 1984; Dijkstra et al., 2012).

Rock fall fences described in this chapter are also suited to the containment of debris flows (Bichler et al., 2012; Bugnion et al., 2012). That is, the strength and flexibility of the fence has the capacity to absorb the impact energy of the flow, and the open fabric of the net allows rapid drainage of the water content. Removal of the water significantly reduces the fluidity and mobility of the debris with the result that the solid material will be contained by the net (Figure 10.21). As the result of extensive full-scale testing and computer modeling of debris flow barriers by manufacturers of these systems, proprietary programs have been developed for barrier design. In the application of these programs, typical design input parameters are

- Flow type—granular or mud
- Density—specific weight and water content
- Volume of event
- Peak discharge—flow rate
- Factor of safety
- Retention volume
- Dimensions of barrier—height, base, and top width, distance to next barrier upslope
- Slopes of upslope stream channel, and retained debris

Figure 10.21 Wire-mesh fence containing debris flow material.

- Impact velocity
- Type of containment fence

Using these input parameters, the software calculates the capacity of the chosen containment fence and determines whether it has adequate capacity in terms of both impact energy and retained volume. The input parameters can be adjusted until an appropriate fence has been selected.

Chapter 11

Rock Fall Protection II—Rock Sheds

Reinforced concrete rock sheds have been developed to provide a highly reliable level of protection on major transportation routes, and at tunnel portals. Rock sheds are used extensively in Japan and Europe; Figure 11.1 shows a variety of concrete shed configurations used in Switzerland (Vogel et al., 2009). However, sheds are less common in North America where the traffic volumes are not as high and lower cost protection methods such as barriers and fences are generally accepted.

11.1 TYPES OF ROCK SHEDS

The most common type of rock shed is precast reinforced concrete; Figure 11.2 shows a typical structure in Japan. This shed comprises a cast-in-place concrete retaining wall backfilled with gravel on the mountainside, with precast columns on the valley side and precast roof beams that support a layer of sand to absorb impact energy. Essential features of these structures are their energy-absorbing components that include cushioning material on the roof and flexible elements in columns and roof slabs as discussed in Section 11.2.4.

Reinforced concrete sheds are usually used at locations with frequent, hazardous rock falls and where very reliable protection is required for facilities such as high-traffic-volume highways and high-speed trains where service interruptions cannot be tolerated. The advantages of concrete sheds are that they can be designed to withstand a specific impact energy capacity that can be greater than most types of wire rope fences and barriers (see Figure 10.1). Furthermore, concrete sheds have a long service life and require little maintenance.

The main disadvantage of sheds is their high construction cost. Cost items include complex, precast reinforced concrete beams and slabs, as well cast-in-place concrete foundations and walls. In order to withstand the substantial dead and live loads of the structure, high-capacity foundations are required, particularly on the valley side of the structure. In steep mountainous terrain, the valley-side slope may be unstable and deep foundations such as rock socketed piles may be needed to transfer the loads to stable bedrock. Furthermore, construction of a shed on an active highway or railway will probably require construction during short duration traffic closures ("work windows") with the result that productivity of the work crew will be low.

Where it is not possible to construct adequate foundations on the valley side of the structure, even more costly cantilevered sheds may be required as described in Section 11.3.

The roofs of most sheds are near horizontal, with a slope of about 5° for drainage, since this configuration limits the span of the roof beams and helps maintain a uniform layer of sand. For sheds below steep slopes, the design will be based on direct impact of rock falls landing on the roof at an angle close to 90° resulting in all the impact energy being absorbed by the structure. In contrast, for sheds below flatter slopes, the rock falls will tend to roll

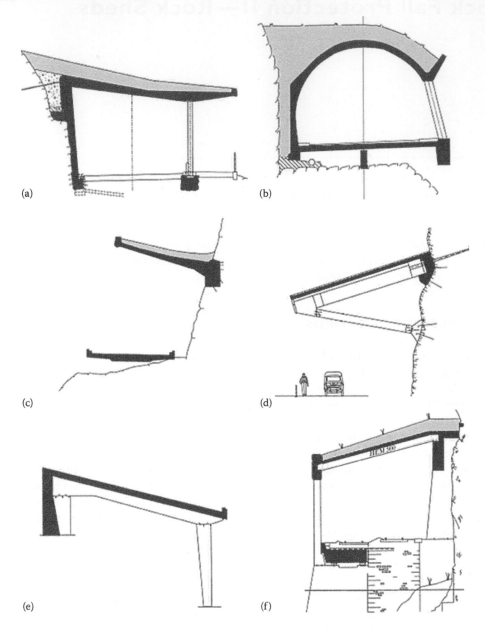

Figure 11.1 Variety of rock fall shed configurations. (From Vogel, T. et al., 2009. Rock fall protection as an integral task. *Structural Engineering International*, SEI 19[3], 304–312, IABSE, Zurich, Switzerland, www.iabse.org.)

across the roof with little impact energy being absorbed by the structure. In some cases it may be possible to construct a shed with a roof at the same angle as the slope such that rocks will roll across the roof with limited impact energy, and it is possible to use a lightweight structure; Section 11.4 discussed sheds that redirect rock falls.

Although steel sheds have superior energy-absorbing properties due to their greater flexibility than concrete sheds, steel sheds are rarely used due to their higher maintenance costs and shorter service life.

Figure 11.2 Typical precast concrete rock fall shed in Japan (Protec Engineering, Niigata, Japan).

A recent (2013) development for rock fall protection is to construct a canopy using steel-wire mesh for the roof member as described in Section 11.5. The advantage of this type of structure is its comparative lightweight and low cost, and the absence of highly loaded foundations.

11.2 REINFORCED CONCRETE SHEDS

In Japan, research on rock fall impact forces on steel rock fall sheds was first carried out in 1973, and by 1987 more than 2000 full-scale tests had been carried out on pre-stressed concrete (PC) and reinforced concrete (RC) rock fall sheds (Yoshida et al., 1987).

This section describes design and construction principles for concrete sheds based on work in Japan and Switzerland. The details of the structural design of reinforced concrete sheds are beyond the scope of this book.

11.2.1 Energy Absorption—Weight and Transmitted Impact Forces

The functioning of rock fall sheds relates to two forces generated by the rock fall impact—the *weight impact force* and the *transmitted impact force* (Figure 11.3). For a rock impacting the roof of a shed, the weight impact force is the product of the mass of the rock and its maximum deceleration as it deforms the cushioning material. The transmitted impact force is the force that is transferred through the cushioning layer into the structure.

When a rock impacts the sand cushion, the sand first consolidates by plastic deformation, and then the rock penetrates the sand when the induced shear stress exceeds the shear strength of the consolidated cushion material. The rock penetrates the cushion material until the energy of the falling rock is completely absorbed by consolidation and shearing of the sand. The transmitted force is determined by integration of the transmitted pressure over the area that it acts on the roof. The transmitted impact force differs from the weight

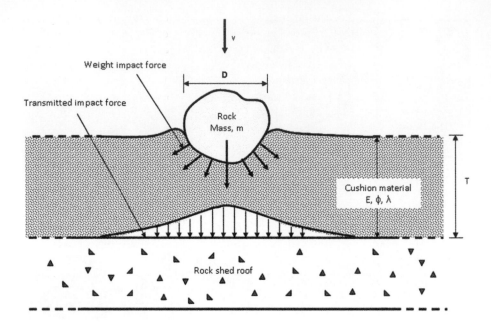

Figure 11.3 Transfer of force from falling rock—weight impact force, through cushioning material into roof of shed—transmitted impact force.

impact force because the cushioning material on the roof both absorbs energy due to plastic deformation, and distributes it over a finite area of the roof that is larger than the area of the impact.

While the cushioning material absorbs the impact energy, the shed structure may amplify the force as the result of dynamic forces being reflected within the roof beams. Full-scale testing of sheds shows that for rigid structures such as concrete sheds, the transmitted impact force may be between 1.5 and 2 times the weight impact force, but there is no simple relationship between these two forces because of the complex dynamic behavior of the cushioning layer and underlying roof beams.

One of the objectives of full-scale rock fall tests with a variety of cushioning materials and their thickness has been to clarify the relationship between the weight impact force and the transmitted impact force (Yoshida et al., 1987). Full-size and model tests, as well as structural analysis, show that the structural characteristics of rock fall sheds have a significant influence on the magnitude of the transmitted impact force (Masuya et al., 1987; Yokoyama et al., 1993). That is, the more flexible the structure, the lower the transmitted force.

11.2.2 Properties of Cushioning Layer

The required properties of cushioning material on the shed roof should be to absorb energy due to plastic compression and distribute the weight impact force from the small area at the point of impact into a wider area on the shed roof. The material should also be inexpensive, have a low density to minimize the dead load, be able to resist the impact forces without damage, and be long lasting.

Sand is generally used as the cushioning material for rock fall sheds because of its low cost, long life, and reasonable energy-absorbing properties. The disadvantage of sand is its weight, and if several meters of sand cushion are used, the shed design is governed more by the dead load of the sand than by the impact force.

In Japan, Styrofoam has been used instead of sand as the cushioning material for sheds subject to severe rock fall impact loads because of its low-density and good energy-absorbing properties (Mamaghani et al., 1999). A comparison between the relative transmitted impact forces on the concrete roof slab generated by a 30 kN (3.5 tonf) weight for sand and Styrofoam cushioning materials is shown on Figure 11.4 where the area under the force-deformation curve equals the impact energy absorbed due to deformation. These results illustrate that Styrofoam, which begins to absorb energy as soon as impact occurs, is more effective than sand in absorbing and dissipating the impact force. However, the tests also showed that unreinforced Styrofoam shatters when impacted, and it is necessary to reinforce the Styrofoam with polypropylene straps, which distribute the impact force over a wider area. Disadvantages of Styrofoam are its high cost compared to sand and its breakdown on exposure to ultraviolet light.

Test results on Styrofoam for very high impact loads have been used to develop a relationship between the magnitude of the transmitted impact force and the thickness of the cushioning layer. Because Styrofoam is homogeneous, it has been found that this formula is applicable to a wide range of impact loads (Yoshida et al., 1991b). One existing rock shed has been retrofitted with a 10 m (33 ft) layer of Styrofoam to provide protection against very large rocks.

The load-deformation test results for sand, Styrofoam, and rubber tires are shown in Figure 11.4, and their cost-effectiveness for rock shed protective layers are as follows:

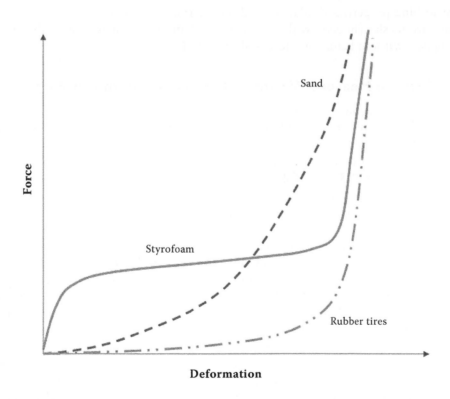

Figure 11.4 Relationship between transmitted impact force and deformation for three cushioning materials. (From Yoshida, H. et al. 2007. Rock fall sheds—application of Japanese designs in North America. *Proc. 1st. North American Landslide Conference*, Vail, CO., AEG Special Publication No. 22., ed. Turner, A. K., Schuster, R. L., pp. 179–196.)

Rubber tires: A pile of rubber tires contains considerable air space, and therefore they are readily crushed when loaded. Once the tires are crushed and the rubber elements are in contact, the transmission force increases rapidly as the rubber is compressed. As a result of this behavior, the energy-absorbing properties of rubber tires is generally not adequate for rock shed construction.

Sand: The force-deformation curve of sand is parabolic, with only a small amount of energy absorption in the initial deformation range. However, the force rapidly increases as the deformation increases. This behavior is due to the sand being loose, so that initial consolidation is necessary before it begins to absorb energy. The density of loose sand is about $20 \text{ kN} \cdot \text{m}^{-3}$ ($130 \text{ lbf} \cdot \text{ft}^{-3}$).

Styrofoam: A layer of Styrofoam provides effective energy absorption due to its stiffness. That is, for small strains in the range of about 5%, the force increases rapidly, and then only increases slowly with increasing deformation. The majority of the energy absorption occurs between 5% and 70% deformation as the Styrofoam deforms plastically. However, when the strain reaches 70%, the Styrofoam is almost fully compressed, and there is a sudden increase in the force. The density of Styrofoam is $0.16 \text{ kN} \cdot \text{m}^{-3}$ ($1 \text{ lbf} \cdot \text{ft}^{-3}$).

Glass granules: An alternative protective material is cellular glass fill made up of 10 to 50 mm (0.5 to 2 in.) diameter glass granules contained in geofabric and steel-mesh cylinders about 1.5 m (5 ft) in diameter (Geobrugg, 2010). Advantages of glass granules are their light weight of $2.5 \text{ kN} \cdot \text{m}^{-3}$ ($15 \text{ lbf} \cdot \text{ft}^{-3}$), long-term durability, and energy-absorbing properties similar to sand. Furthermore, if the granular glass is contained in wire-mesh cylinders, with a diameter of 1 m (3.3 ft), it is able to absorb multiple impacts without being displaced and dispersed.

11.2.3 Tests to Measure Weight and Transmitted Impact Forces

The two previous sections describe the energy absorbance of protective layers and the concepts of the weight impact force and the transmitted impact force. Values for these two forces have been obtained by full-scale testing using equipment such as that shown in Figure 11.5. The test data have been used to calibrate analytical models of sheds. Features of the test equipment in Figure 11.5 are as follows.

The falling weight contains an accelerometer that measures the deceleration of the weight as it impacts the cushion material. The weight impact force is defined as the mass multiplied by the maximum value of the sampled deceleration during its impact with the sand.

The force that is transmitted into the roof of the shed is measured with an array of earth pressure cells located on the upper side of the roof at the base of the cushion layer and arranged at constant intervals from the impact point. The earth pressure cells measure the pressure per unit area at each sampling position, and these measurements are integrated to find the total transmitted force, assuming that the pressure is distributed axisymmetrically about the impact point.

Calculation of the transmitted impact force requires integration of the maximum measured pressure values over the sampling time. It is found that reliable values for the weight and transmitted impact forces require sampling rates of up to 2,000 Hz to ensure that peak values are measured because the transient forces have high frequencies that may not be identified at lower sampling rates. It is also necessary that precise measurements are made of low stresses at large distances from the impact point in order to accurately integrate the full transmission force.

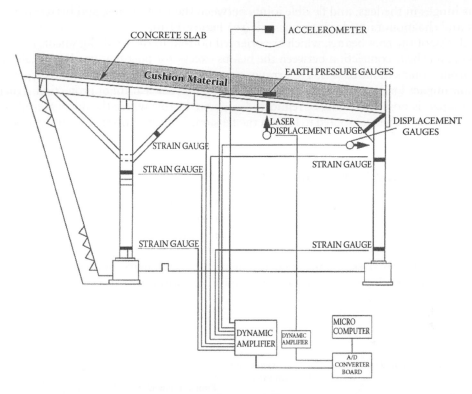

Figure 11.5 Instrumented shed to measure weight impact force and transmitted impact force of test rock falls (Yoshida et al., 2007).

The performance of the shed structure is measured with strain gauges and displacement gauges as shown in Figure 11.5. This information is used to determine the required dimensions and strength of each structural member.

11.2.4 Shed Design—Flexibility and Cushioning

The results of tests such as those illustrated in Figure 11.5 show that the two factors that most influence the magnitude of the transmitted impact force are the flexibility of the structure and the characteristics of the cushion layer.

Although the cushioning layer absorbs and distributes the impact force, the transmitted impact force is generally greater than the weight impact force due to the interaction between the dynamic impact load and the roof beams. For example, for a concrete shed protected with a 1 m (3.3 ft) thick sand cushion, the transmitted impact force can be 1.5 to 2 times the weight impact force. Furthermore, the maximum energy absorption occurs where the thickness of the cushion layer is equal to the diameter of the rock fall, and any additional thickness just adds to the dead weight of the structure without increasing the energy absorption. The tests also show that while Styrofoam is more effective than sand in absorbing the impact energy, the transmitted impact energy is still greater than the weight impact energy. For example, for a 3.5 m (12 ft) thick layer of Styrofoam impacted by a 5,000 kg (11,000 lb) mass falling from a height of 20 m (65 ft), transmitted impact force was 1.2 times the weight impact force (Mamaghani et al., 1999).

With respect to the flexibility of the structure, the transmitted impact force is reduced for a steel shed compared to a relatively rigid concrete shed, and the flexibility increases with the span of the roof. In addition, flexible components can be built into concrete structures such as hinges in the legs, and flexible joints between the roof beams, and between the roof beams and the mountainside retaining wall (see Figure 11.6).

Flexibility of the roof beams, which are oriented normal to the road alignment, is achieved by having no shear connection between the beams except for post-tensioned steel cables running through ducts in the beams. The arrangement produces both flexibility and dispersion of the point impact load over a larger area of the roof. Another feature of protection structures built in Japan is post-tensioning of reinforced concrete beams where the ducts are filled with grease, rather than being grouted to lock-in the tension. The tensioned cables and grease-filled ducts produce an effective combination of strength and flexibility (Protec Engineering, 2012).

11.2.5 Typical Rock Shed Design

Figure 11.6 shows cross-section and elevation views of a rock fall shed designed for a single-track railway using Japanese design procedures (Yoshida et al., 2007). The dimensions and layout of the shed are dictated by the following site conditions:

- **Impact energy**—Design impact energy defined by rock mass and velocity.
- **Clearance envelope**—The inside dimensions conform to the dynamic clearance envelope for the trains.

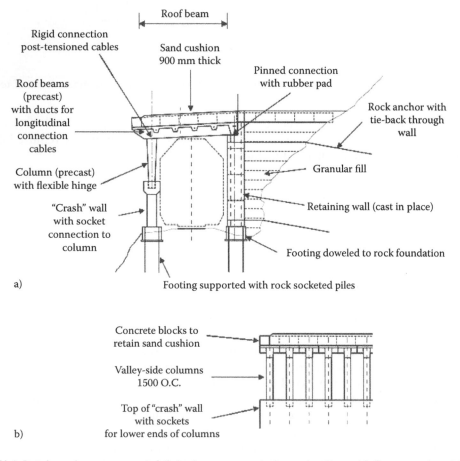

Figure 11.6 Reinforced concrete rock fall shed to protect single track railway. (a) Cross-section, (b) elevation. (Courtesy of Protec Engineering, Japan.)

- **Retaining wall**—The space between the rock face and the mountainside wall of the shed is filled with gravel that provides protection for the shed from rock falls with trajectories close to the rock face. The gravel is contained by the cast-in-place wall on the mountainside of the shed and two end walls.
- **Roof gradient**—The roof grade of 5% provides for drainage of the cushioning sand. In general, a horizontal roof minimizes the span width and facilitates the maintenance of a uniform layer of sand. Sheds with sloping roofs that redirect rock falls are discussed in Section 11.4 below.
- **Valley-side foundation**—The slope below the track is formed by loose fill that is not able to support the foundation load of the shed. Therefore, the foundation is supported with rock-socketed piers drilled through the fill into bedrock (Wyllie, 1999). The load on the foundation comprises the dead load of the shed, the equivalent static load of rock fall impacts and the horizontal force applied to the crash wall (see below).
- **Valley-side columns**—The columns are precast structures at a longitudinal spacing of 1,500 mm (60 in.) to match the width of roof beams. The lower end of each column, just above the socket, incorporates a hinge made up of galvanized steel bars bolted to the reinforcing steel cages above and below the hinge. Flexibility in the hinge is provided by a layer of synthetic rubber set into the concrete.

  In the top of each beam, prestressing cables create a rigid connection between the columns and the roof beams (see below).

  A requirement for this railway application was a "crash barrier" on the valley side of the shed that would allow an impact by a derailed train without damage to the structure.
- **Roof beams**—The roof beams are precast structures that are arranged transverse to the track alignment. The outer ends of the beams incorporate holes for the poststressing cables in the columns. The beams are rigidly connected to the tops of valley-side columns by tensioning and locking off the cables on the upper surface of the beams. The roof beams themselves are joined together with a set of post-tensioned cables that run in ducts arranged parallel to the track.
- **Mountainside wall**—The mountainside wall is a continuous, cast-in-place structure that both supports the roof beams and retains the fill. The wall incorporates rock anchors in the rock face and tie-backs through the wall to resist the fill load and water pressure that develop behind the wall. The connection between the top of the wall and the roof beams is a hinged joint comprising steel pins grouted into holes in the concrete, with a neoprene rubber pad between the wall and the roof beam.

The use of precast elements for most of the structure facilitates construction, particularly where the site is an active railway where columns and beams need to be erected quickly during short work "windows" between trains. However, construction of a shed of this type, even at a site with no traffic, would be expensive with forming and pouring the valley-side wall, the need for cranes to lift the columns and beam into place, and the placement and distribution of sand on the roof.

11.2.6 Static Equivalent Force

Tests such as those using the equipment shown in Figure 11.5 that provide actual values of the transmitted impact forces in the shed show considerable scatter in the values of the transmitted force. Therefore, for the purpose of shed design, equations for a static force that is equivalent to the dynamic impact load have been developed in Japan and Switzerland. The

Japanese equation is based on Hertzian impact theory for impact between elastic spheres, while the Swiss equation is based on impact tests. These two equations are discussed below.

Japan: The static equivalent force P (kN) for a mass m (kg) falling from a height H (m) is given by (Japanese Road Association, 2000)

$$P = 0.02 \, (m \cdot g)^{0.67} \cdot \lambda^{0.4} \cdot H^{0.6} \cdot \beta \tag{11.1}$$

where λ is the Lamé parameter for the cushion material with modulus of deformation E and Poisson's ratio v, given by:

$$\lambda = \frac{E \cdot v}{(1+v)(1-2v)} \tag{11.2}$$

Approximate values of the Lamé parameter for sand commonly used on rock fall sheds are

$\lambda = 1{,}000 \text{ kN} \cdot \text{m}^{-2}$ for very soft sand
$\lambda = 3{,}000$ to $5{,}000 \text{ kN} \cdot \text{m}^{-2}$ for firm sand
$\lambda = 10{,}000 \text{ kN} \cdot \text{m}^{-2}$ for stiff sand

In Equation (11.1), the factor β defines the relationship between the thickness of the cushioning layer (T) and the diameter of the impacting rock (D) given by

$$\beta = \left(\frac{T}{D}\right)^{-0.58} \tag{11.3}$$

It is common in Japan to use a sand cushion with thickness, $T = 0.9$ m (3 ft) that balances the requirements to have a cushion thick enough to absorb impact energy and to limit the weight of the cushion that has to be supported by the structure. Evaluation of Equation (11.3) shows that the value of the parameter β does not decrease significantly as the ratio T/D is increased by increasing the sand thickness.

Switzerland: The following equation for the static equivalent force P has been developed in Switzerland (Schellenberg et al., 2009; Jacquemoud et al., 1999; Labiouse et al., 1996) using the same units as Equation (11.1):

$$P = C \cdot 0.028 \cdot T^{-0.5} \cdot D^{0.7} \cdot E^{0.4} \tan\varphi \left(\frac{m \cdot V^2}{2}\right)^{0.6} \tag{11.4}$$

where C is a coefficient to account for ductile ($C = 0.4$) or brittle ($C = 1.2$) failure, E and φ are the modulus of deformation and friction angle, respectively, of the sand cushion material, and V is the impact velocity.

Worked Example 11A—static impact force calculations: The following calculations compare the values for the static equivalent force P given by Equations (11.1) and (11.4).

For a spherical rock mass with diameter 0.5 m and density 2,650 kg $\cdot$ m^{-3}, the mass m is 175 kg. The cushioning material is loose sand with Lamé parameter $\lambda = 1{,}000$ kN $\cdot$ m^{-2}, modulus,

E = 10,000 kN · m^{-2} and friction angle φ = 40°, and the thickness of the cushion is 0.6 m. The shed is designed for ductile failure, so C = 0.4. If the rock falls from a height of 50 m, the impact velocity will be 31 m · s^{-1}, and the two values for the static equivalent force are:

Japan—P = 426 kN; Switzerland—P = 268 kN

The Japanese equation, which is a theoretical relationship, is more conservative in many conditions than the Swiss equation, which is based on test results.

11.3 CANTILEVERED STRUCTURES

For site conditions where the foundations on the outside of the road or railway are very poor, it may be necessary to construct a cantilevered shed such as those shown in Figure 11.1(c)(f). Another possible reason for using this type of structure is where it is necessary to have clearance for maintenance equipment, for example, to operate on the shoulder of the road or track. For a cantilevered shed, cost savings are achieved by eliminating the valleyside columns and foundations. However, cantilevered sheds require high strength rock in which to install anchors and foundations, and/or enough space at the base of the rock slope to build supporting structures.

The two examples of cantilevered sheds shown in Figure 11.7 illustrate construction methods. The shed in Figure 11.7(a) has been constructed mainly to deflect snow avalanches

(a)

(b)

Figure 11.7 Examples of cantilevered rock sheds. (a) Cantilevered concrete shed used to deflect rock falls and snow avalanches over railway track (courtesy of Canadian Pacific Railway); (b) "rock keeper" structure. (Courtesy of Protec Engineering, Japan.)

over the track, but has also been effective in protection against rock falls. The rock face just above the track at a slope angle of about 45° has allowed the roof to be configured such that the length of the beams to the right of the support columns is longer than the cantilevered section to the left of the support column. This roof geometry has limited the uplift dead load along the mountain edge of the roof so that the rock bolts anchoring the roof are mainly required to resist the live load of avalanches.

The structure in Figure 11.7(b) is termed a "rock keeper" and is commonly used in Japan. The rock keeper is in effect an elevated ditch with a layer of sand on the concrete to provide protection. This structure will contain rock falls and must withstand the entire impact energy, compared to the cantilevered shed which redirects the fall and need only absorb part of the impact energy.

11.4 SHEDS WITH SLOPING ROOFS

The type of shed shown in Figures 11.2 and 11.6 with an approximately horizontal roof is designed for steep trajectories where the impact is approximately normal to the roof. However, in some locations, the slope geometry may lend itself to the construction of a shed that redirects the rock across the roof with low-impact forces on the structure.

Figure 11.8 shows a shed in the base of a gully with the slope of the roof close to the slope of the rock face in the gully. This configuration of the shed allowed a lightweight structure to be built where the rocks roll and slide across the roof with negligible impact. The angle of the roof is steep enough that rock does not accumulate on the roof, which limits the dead load.

The disadvantage of the shed configuration shown in Figure 11.7 is that the upslope width of the roof is about five times the width of a horizontal roof that just spans the railway track. Also, the slope of the roof is too steep for the use of a sand protection layer; near horizontal roofs are required to maintain a uniform layer of sand on the roof.

Figure 11.8 Rock shed with sloping roof that redirects rock, with low-impact forces on the structure. (Courtesy of Canadian National Railway.)

11.5 WIRE-MESH CANOPIES

An alternative to the various reinforced concrete sheds shown in Figure 11.1 is to use a canopy constructed with wire mesh, supported on steel posts and cables as shown in Figure 11.9.

Canopies function by redirecting rock falls such that only a portion of the impact energy is absorbed by the canopy, with the balance of the energy being retained in the moving rock as it passes over the lower edge of the structure. That is, canopies function similarly to fences inclined upslope and hanging nets discussed in Chapter 9 where rock falls are redirected, but not stopped by the structure. The actual reduction in velocity depends on the angle of impact, with rocks that fall vertically on to the roof being stopped so that all the impact energy is absorbed by the canopy. In contrast, for oblique impacts on steeply sloping roofs, little reduction in velocity occurs and only a small portion of the impact energy is absorbed by the canopy. Therefore, it is beneficial that canopies are designed, if possible, with the roof inclined close to the angle of the slope face.

Figure 11.10 shows an example of a three-dimensional analysis carried out to design a canopy. The purpose of the analysis was to determine the forces generated in all the canopy components, to study the behavior of the rock falls with respect to deflection of the net relative to the clearance envelope and whether the rock would roll off the roof and not accumulate on the mesh.

The following is a discussion of some of the design features of wire-net canopies.

Structure flexibility—The wire mesh forming the roof and the braking elements in the support cables produces a flexible structure that can absorb energy by deformation in comparison with more rigid concrete structures. This flexibility, in combination with the

Figure 11.9 Wire mesh and Ringnet canopy. (Courtesy of Geobrugg, Switzerland.)

inclined roof that redirects falls, allows a relatively light structure to have the same impact energy capacity as a heavier concrete shed.

Clearance envelope—Canopy dimensions are determined by the clearance envelope of trucks or railway cars that are being protected. However, allowance must be made for the deformation of the roof that will occur during impact by a rock fall such that the vehicles or trains do not strike rock falls contained in the deflected net. This will require that the canopy be located at a height so that the deflected net is above the clearance envelope.

Self-cleaning roof—It is desirable that the roof is self-cleaning, with rock not accumulating on the mesh, in order to minimize maintenance; this will require that the roof is at a uniform slope at an angle of 30° to 40°. Self-cleaning action also depends on the impact angle and rotation of the rock falls.

Anchors and foundations—A significant advantage of wire-mesh canopies compared to concrete sheds is that highly loaded foundations are not needed on the valley side of the structure where stability conditions may be marginal. The canopy shown in Figure 11.9 has a series of cables anchored on the outside of the road to provide a vertical restraint on the support beams. However, canopies can be configured with guy wires anchored on the valley-side slope as shown in Figure 11.10, thus eliminating the outside guy wires and anchors. This configuration may be required where outside guys would interfere with operations such as snow clearing or rail maintenance.

Furthermore, the relatively light structure can be suspended on the rock face from the rock bolts.

Anchors to support the beams and guy cables comprise rock bolts, or groups of rock bolts, grouted into holes drilled in the rock face. The diameters of the bolts will usually be in the range of 25 to 35 mm (1 to 1.4 in.), and the depth of embedment in the rock will be depend on the bond strength between the cement grout and the rock in the periphery of the drill hole. Section 10.5.1 discusses the design of rock anchors and the calculation of bond lengths.

Construction considerations—Construction of canopies requires drilling and installing rock anchors into the rock face, and then lifting beams and mesh into place. This work would usually be carried out using a crane, and it will be necessary to match the lift capacity and reach of the crane to the height of the anchors and the weight of the canopy components. For example, on narrow work benches, the extension of the outriggers on the crane may be limited and this, in turn, will limit the lift capacity. In circumstances where drilling for the rock bolts has to be carried out with handheld drills, the maximum drill-hole diameter may be 50 to 75 mm (2 to 3 in.), in which case the diameter of the rock bolts would be limited to 25 mm (1 in.) in order to fully embed the bolts in grout. A number of these bolts

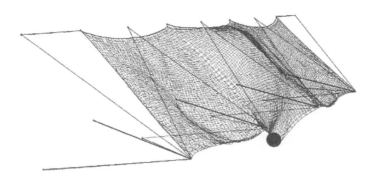

Figure 11.10 Three-dimensional dynamic analysis of wire-mesh canopy using computer program ABACUS to determine forces in canopy components and examine rock fall behavior.

could be installed in a group to provide the necessary load capacity, with the bolts splayed outward to produce an inverted cone of reinforced rock with high pullout resistance.

Another construction consideration is worker safety since canopies (and sheds) would presumably be installed in locations with high rock fall frequency. Precautions for safety that may be taken are to only work during periods of warm, dry weather with low winds when the rock fall hazard is low. Figure 1.3 shows a typical relationship between rock fall frequency and weather conditions. In addition, the platform on which the crew is working could be covered with a steel mesh canopy to protect them from rock falls.

Appendix I: Impact Mechanics— Normal Coefficient of Restitution

This appendix shows the derivation of the equations defining the normal coefficient of restitution in terms of the energy loss during compression and the energy recovery during restitution phases of impact. This is an expansion of the information contained in Section 4.4 of the main text.

The principle of separating the compression and restitution phases of impact can be demonstrated on a normal impulse p_N, relative velocity v $[p_N - v]$ plot as shown in Figure I.1; v is the relative velocity at the impact points. On this plot, the normal velocity changes during impact, starting with a negative value $(-v_{iN})$ at the point of impact, increasing to zero at the point of maximum compression p_c, and finally reaching a positive value (v_{fN}) at the point of separation. Also, the tangential velocity v_T decreases continuously during impact from v_{iT} at the point of impact, to v_{fT} at the point of separation. The change in normal velocity is the result of plastic deformation of the body and slope during impact, while the change in tangential velocity is the result of frictional resistance on the contact surface.

The $[p_N - v]$ plot on Figure I.1 shows the changes in both the normal (v_N) and tangential (v_T) velocity components, and the magnitude of the internal energy of deformation generated during impact. These changes in velocity, as well as energy, can be quantified in terms of the coefficient of restitution, e that has normal (N) and tangential (T) components as follows:

$$e_N = -\frac{v_{fN}}{v_{iN}} \tag{I.1}$$

and

$$e_T = \frac{v_{fT}}{v_{iT}} \tag{I.2}$$

where the subscript "i" refers to the initial velocity at the moment of impact, and the subscript "f" refers to the final velocity at the end of the impact.

The normal impulse, p_N, is defined as the application of a force, F over time:

$$dp_N = m\,(v_N - v_{iN}) = F\,dt$$

or

$$dv_N = \frac{dp_N}{m} \tag{I.3a}$$

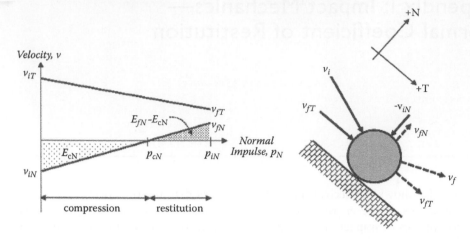

Figure I.1 Relationship between normal impulse p_N and changes in tangential and normal velocities v_T, v_N, and energy during impact.

and the normal impulse p_N, between times $t = i$ and $t = t$ is given by:

$$m\,(v_N - v_{iN}) = \int_i^t F\,dt = p_N \tag{I.3b}$$

The relative normal velocity v_N at any time t during the impact can be obtained by integration of Equation (I.3b), for the initial condition, at the moment of impact $t = i$, and the normal velocity is v_{iN}:

$$v_N = \int_i^t \frac{1}{m} d\,p_N \tag{I.4}$$

$$= \frac{1}{m}[p_N - p_{iN}]$$

$$v_N = v_{iN} + \frac{p_N}{m} \tag{I.5}$$

where $v_{iN} < 0$.

The impacting normal velocity is negative because, as shown in Figure I.1, the positive normal axis is in the direction away from the point of impact.

Equation (I.5) shows that the relative velocity is a linear function of the normal impulse, which is expressed as a straight line on Figure I.1.

Equation (I.5) can be used to find the impulse at maximum compression (p_{cN}). At the point of maximum compression ($t = c$), the normal velocity is momentarily equal to zero and the corresponding normal impulse has a value p_{cN} given by the following equation:

$$0 = v_{iN} + \frac{p_{cN}}{m}$$

and

$$p_{cN} = -m \cdot v_{iN} \tag{I.6}$$

At the end of the impact ($t = f$), the final normal velocity is v_{fN} and the final normal impulse (p_{fN}) can also be found from Equation (I.5),

$$p_{fN} = (m \cdot v_f - m \cdot v_i) \tag{I.7}$$

With respect to the energy of deformation, the triangular area E_{cN} on Figure I.1 represents the kinetic energy of normal motion that is absorbed in compressing the deformable region, while triangular area $(E_{fN} - E_{cN})$ represents the elastic strain energy recovered during restitution that drives the body from the slope.

Expressions for these two energy changes are

<u>Compression:</u>

$$E_{cN} = \int_0^{p_{cN}} v_N \, dp = \int_0^{p_{cN}} \left(v_{iN} + \frac{p_N}{m} \right) dp \tag{I.8}$$

$$= \left(v_{iN} \cdot p_{cN} + \frac{p_{cN}^2}{2\,m} \right)$$

$$= -\frac{1}{2} m \cdot v_{iN}^2$$

Equation (I.8) shows that all the normal impact kinetic energy is absorbed up to the point of maximum compression.

<u>Restitution:</u>

$$(E_{fN} - E_{cN}) = \int_{p_{cN}}^{p_{fN}} \left(v_{iN} + \frac{p_N}{m} \right) dp = \left[v_{iN} \cdot p_N + \frac{p_N^2}{2 \cdot m} \right]_{p_{cN}}^{p_{fN}}$$

$$= \left(v_{iN} + \frac{p_{fN}^2}{2 \cdot m} - \left(v_{iN} \cdot p_{cN} + \frac{p_{cN}^2}{2 \cdot m} \right) \right)$$

$$= v_{iN} + \frac{p_{fN}^2}{2 \cdot m} - \left(-m \cdot v_{iN}^2 + \frac{m \cdot v_{iN}^2}{2} \right)$$

where $p_{cN}^2 = m^2 \cdot v_{iN}^2$

$$(E_{fN} - E_{cN}) = v_{iN} \cdot p_{fN} + \frac{p_{fN}^2}{2\,m} + \frac{m \cdot v_{iN}^2}{2}$$

where

$$v_{iN} = \frac{-m \cdot v_{iN}^2}{p_{cN}}$$

and

$$\frac{1}{m} = \frac{m \cdot v_{iN}^2}{p_{cN}^2}$$

$$(E_{fN} - E_{cN}) = \frac{m \cdot v_{iN}^2}{2} + \left(-\frac{m \cdot v_{iN}^2}{p_{cN}} \right) p_{fN} + \left(\frac{m \cdot v_{iN}^2}{2 \cdot p_{cN}^2} \right) p_{fN}^2 \tag{I.9}$$

$$= \frac{m \cdot v_{iN}^2}{2} \left(1 - 2 \frac{p_{fN}}{p_{cn}} + \frac{p_{fN}^2}{p_{cN}^2} \right)$$

$$= \frac{1}{2} m \cdot v_{iN}^2 \left(\frac{p_{fN}}{p_{cN}} - 1 \right)^2$$

where $v_{iN} < 0$.

The expressions in Equations (I.8) and (I.9) for the partially irreversible changes in kinetic energy of normal motion that occur during impact can be used to define the normal coefficient of restitution, e_N, as follows:

$$e_N^2 = -\frac{E_N(p_{fN}) - E_N(p_{cN})}{E_N(p_{cN})} \tag{I.10}$$

This definition of the coefficient of restitution in terms of energy separates the energy loss due to plastic compression and hysteresis of the contact forces from that due to friction and slip between the colliding bodies.

The relationships shown in Equations (I.8), (I.9), and (I.10) can be combined, for normal impact, to find the following expression for the normal coefficient of restitution in terms of the normal impulses at maximum compression (p_{cN}) and at the completion of the impact (p_{fN}):

$$e_N^2 = \left(\frac{p_{fN}}{p_{cN}} - 1 \right)^2 \tag{I.11}$$

Substitution of the expressions for p_{cN} and p_{fN} in Equations (I.6) and (I.7) into Equation (I.11), yields the following expression relating impulse to the coefficient of restitution:

$$p_{fN} = -m \cdot v_{iN} (1 + e_N) = p_{cN} (1 + e_N) \tag{I.12}$$

and for the normal coefficient of restitution,

$$e_N = -\frac{v_{fN}}{v_{iN}} \tag{I.1}$$

$$e_N = \frac{(p_{fN} - p_{cN})}{p_{cN}} \tag{I.13}$$

As shown in Figure I.1 and expressed in Equation (I.1), the normal coefficient of restitution is the ratio of final normal velocity to the impact normal velocity, and is also the square root of the negative ratio of the energy recovered during the restitution phase of the impact to the energy lost during the compression phase (Equation I.10).

For rough bodies where slip occurs at the contact point, but the direction of slip is constant, the two expressions for the coefficient of restitution in Equations (I.1) and (I.13) are equivalent.

Appendix II: Impact Mechanics— Impact of Rough, Rotating Bodies

This appendix shows the derivation of equations defining the final translational and rotational velocities due to impact of a rough, rotating body. This is an expansion of the information contained in Section 4.6.2.

II.I EQUATIONS OF RELATIVE MOTION

For a rotating body impacting a slope at an oblique angle, the body will have both translational v, and angular ω, velocities. The relative velocities at the impact points, expressed as normal and tangential components are, v_N and v_T (Figure II.1). At the contact point, equal and opposite forces, F, $-F$, are developed that oppose interpenetration of the body into the slope and give differentials of impulse dp over time dt in the normal and tangential directions that are related by:

$$dp = F\,dt \tag{II.1}$$

From Newton's second law (see Section 4.1.2), equations of motion for translation of the centre of mass of the body with mass m and velocity V_T are in the normal N and tangential T axes are:

$$dV_N = \frac{dp_N}{m} \tag{II.2}$$

and

$$dV_T = \frac{dp_T}{m} \tag{II.3}$$

and for planar rotation of a body with radius r and radius of gyration k:

$$d\omega = \frac{r}{m\,k^2}\,dp \tag{II.4}$$

The relative velocity at the contact point, v, is the difference between the velocity at the center of the mass, V, and the peripheral velocity $r \cdot \omega$. The relative velocity havs components v_N and v_T in the normal and tangential directions.

Based on these relationships, the planar relative velocity changes are given by:

$$\left\{ \begin{array}{c} dv_T \\ dv_N \end{array} \right\} = m^{-1} \left[\begin{array}{cc} \beta_1 & -\beta_2 \\ -\beta_2 & \beta_3 \end{array} \right] \left\{ \begin{array}{c} dp_T \\ dp_N \end{array} \right\} \tag{II.5}$$

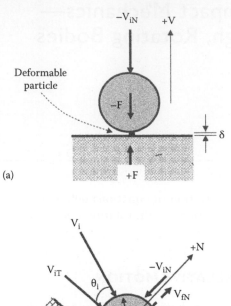

Figure II.1 Impact mechanics principles for two-dimensional (planar) motion: (a) forces generated at contact point during normal impact; (b) impact of rough, rotating sphere on a slope, V = velocity at center of mass, v = relative velocity at impact point.

from which the tangential and normal components of the velocity changes are

$$dv_T = \frac{(\beta_1 \, dp_T - \beta_2 \, dp_N)}{m} \tag{II.6a}$$

and

$$dv_N = \frac{(-\beta_2 \, dp_T + \beta_3 \, dp_N)}{m} \tag{II.6b}$$

where β_1, β_2, and β_3 are inertial coefficients. The definition of the inertial coefficients is shown in Figure II.2 where a body with mass M impacts the slope with mass M'. A local axis system is set up at the impact point with the tangential (T) axis parallel to the slope, and the normal (N) axis normal to the slope with positive away from the slope. The dimensions of the body relative to its center of mass and the point of impact are defined by tangential and normal radii, r_T and r_N and the radius of gyration k_r. If the body is rotating about a fixed z axis, the components of the moments of inertia for plane motion are defined by the tensor, I':

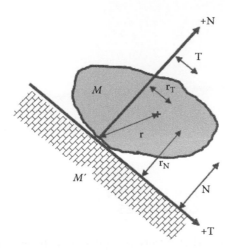

Figure II.2 Dimensions of rotating, impacting body defining inertial coefficients for plane motion, rotating about z axis through center of gravity (+).

$$I' = \begin{pmatrix} m\,(r_N^2 + z^2) & -m \cdot r_T \cdot r_N \\ -m \cdot r_N \cdot r_T & m\,(z^2 + r_T^2) \end{pmatrix} \tag{II.7}$$

that yields three components of the moment of inertia I:

$$I_{TT} = m\,(r_N + z^2); \quad I_{TN} = -m \cdot r_T \cdot r_N; \quad I_{Tz} = -m\,(r_T^2 + z^2) \tag{II.8}$$

Inertial coefficient β_1 is in relation to the term r_N^2 where $z = 1$:

$$\beta_1 = 1 + \frac{m \cdot r_N^2}{M \cdot k_r^2}$$

Inertial coefficient β_2 is in relation to the term $m \cdot r_T \cdot r_N$:

$$\beta_2 = \frac{m \cdot r_N \cdot r_T}{M \cdot k_r^2} \tag{11.9}$$

Inertial coefficient β_3 is in relation to the term $(z^2 + r_T^2)$:

$$\beta_3 = 1 + \frac{m \cdot r_T^2}{M \cdot k_r^2}$$

where m is the effective mass of the two impacting bodies and is defined as

$$m = (M^{-1} + M'^{-1})^{-1} \tag{II.10}$$

Since the mass of the slope M' is very large, $m = M$, the inertial coefficients are reduced to

$$\left.\begin{aligned}\beta_1 &= 1 + \frac{r_N^2}{k^2}\\[1em]\beta_2 &= \frac{r_N \cdot r_T}{k^2}\\[1em]\beta_3 &= 1 + \frac{r_T^2}{k^2}\end{aligned}\right\} \tag{II.11}$$

Equations (II.11) for the inertial coefficients can be applied to any body shape with radius r and radius of gyration k.

II.2 EQUATIONS OF PLANAR MOTION FOR IMPACT OF ROUGH BODIES

For usual rock fall conditions where friction is generated at the impact point, the body may initially slip from the point of impact until a normal impulse value p_{sN}, and then roll, during which time the relative tangential velocity at the contact point v_T, is zero (Figure II.3). According to Coulomb's law, if the coefficient of friction is μ, then the body will slip at the contact point when the relationship between the tangential and normal impulses is:

$$p_T = -\mu \cdot p_N \tag{II.12}$$

From Equation (II.5), the change in relative velocity components during the period of slip in terms of the normal impulse p_N, are:

$$\left\{\begin{aligned} dv_T \\ dv_N \end{aligned}\right\} = m^{-1} \begin{bmatrix} \beta_1 & -\beta_2 \\ -\beta_2 & \beta_3 \end{bmatrix} \left\{\begin{aligned} -\mu\, dp_N \\ dp_N \end{aligned}\right\} \tag{II.13}$$

and the equations for the change in relative velocity components are given by:

$$dv_T = \frac{(-\mu \cdot \beta_1 - \beta_2)\, dp_N}{m} \tag{II.14a}$$

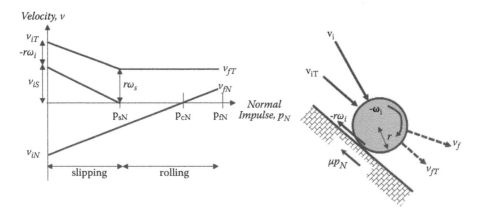

Figure II.3 Changes in rotational (ω) and slip (v_s) velocities during impact, and transition from slip to rolling mode when $v_s = 0$; for negative angular velocity: $v_s = (v_T - r \cdot \omega)$.

and

$$dv_N = \frac{(\mu \cdot \beta_2 + \beta_3)\ dp_N}{m} \tag{II.14b}$$

Integration of these equations for dv_T and dv_N for the period $t = i$ to $t = v_s$ gives components of relative velocity for any normal impulse p_N, during the period of slip

$$v_T = v_{iT} - \frac{(\beta_2 + \mu \cdot \beta_1)}{m} p_N \tag{II.15a}$$

and

$$v_N = v_{iN} + \frac{(\beta_3 + \mu \cdot \beta_2)}{m} p_N \tag{II.15b}$$

Equations (II.15a, b) show that the relationship between v and p_N is linear with the gradient being proportional to $(1/m)$. The equations also show that tangential velocity decreases throughout the impact from the impact value v_{iT}, while the normal velocity increases from an initial negative value to a final positive value as it leaves the slope.

At the point of maximum compression during impact, the normal relative velocity is zero and Equation (II.15b) gives the following expression for the normal impulse p_{cN}

$$v_N = 0 = v_{iN} + \frac{(\beta_3 + \mu \cdot \beta_2)}{m} p_{cN}$$

and

$$p_{cN} = \frac{-m \cdot v_{iN}}{(\beta_3 + \mu \cdot \beta_2)} \tag{II.16}$$

From Appendix I, Equation (I.12), the relationship between the impulses at maximum compression and the completion of the impact is:

$$p_{fN} = -m \cdot v_{iN}(1 + e_N) = p_{cN}(1 + e_N) \tag{11.17}$$

from which it is possible to develop, for a rough, rotating body, an expression for the impulse at the end of the impact, p_{fN}

$$p_{fN} = \frac{-(1 + e_N)\ m \cdot v_{iN}}{(\beta_3 + \mu \cdot \beta_2)} \tag{II.18}$$

II.3 EQUATIONS OF MOTION FOR TRANSLATING AND ROTATING BODIES—FINAL VELOCITIES

The angular velocity ω can be incorporated into the impact mechanics equations (II.5) to (II.11) as follows.

For a collinear impact of a sphere with radius r, Figure II.3 shows that, $r_T = 0$ and $r_N = r$, and Equations (II.11) are further reduced to:

$$\beta_1 = 1 + \frac{r^2}{k^2}; \quad \beta_2 = 0; \quad \beta_3 = 1$$

and Equations (II.15a,b) for the tangential and normal relative velocity components simplify to:

$$v_T = v_{iT} - \frac{(\mu \cdot \beta_1)}{m} p_N \qquad (II.19a)$$

$$v_N = v_{iN} + \frac{p_N}{m} \qquad (II.19b)$$

For the impacting sphere that is a rigid body, the translational velocity components at the center of mass are designated by V_T and V_N and the differential equations for the changes in velocity components in terms of the normal impulse are:

$$dV_T = \frac{-\mu}{m} dp_N; \quad dV_N = dp_N; \quad d\omega = \frac{-\mu \cdot r}{m \cdot k^2} dp_N \qquad (II.20)$$

The components of relative velocity at the contact point for collinear impact are:

$$v_T = V_T + r \ \omega; \quad v_N = V_N \qquad (II.21)$$

The equations for the relative velocity at the point of impact in terms of the normal impulse are obtained by integration, for the initial conditions $(v_{iT} = V_{iT} + r \cdot \omega_i)$ and $(v_{iN} = V_{iN})$ as follows:
Tangential velocity:

$$v_T = v_{iT} + \int dv_T + r \int d\omega \qquad (II.22)$$

$$v_T = v_{iT} + \int \frac{-\mu}{m} dp_N + r \int \frac{-\mu \cdot r}{m \cdot k^2} dp_N$$

$$v_T = v_{iT} - \frac{\mu}{m} \left(1 + \frac{r^2}{k^2}\right) p_N$$

Normal velocity:

$$v_N = v_{iN} + \frac{1}{m} \int dp_N \qquad (II.23)$$

$$v_N = v_{iN} + \frac{p_N}{m}$$

The transition from compression to restitution occurs at the point of maximum compression when $(v_{cN} = 0)$ and $(p_{cN} = -m \cdot v_{iN})$. Using these relationships, the equations for the relative tangential and normal relative velocities can be expressed as dimensionless ratios as follows:

$$\frac{v_T}{v_{iN}} = \frac{v_{iT}}{v_{iN}} + \mu\left(1 + \frac{r^2}{k^2}\right)\frac{p_N}{p_{cN}}$$ (II.24)

and

$$\frac{V_N}{V_{iN}} = 1 - \frac{p_N}{p_{cN}}$$ (II.25)

When slip stops prior to separation $(p_s < p_f)$, the relative tangential velocity $v_{sT} = 0$, and Equation (II.24) can be solved to determine an expression for the impulse ratio at time $t = s$:

$$\frac{p_{sN}}{p_{cN}} = -\frac{v_{iT}/v_{iN}}{\mu(1 + r^2/k^2)} = -\frac{(V_{iT} + r \cdot \omega_i)}{V_{iN}\,\mu(1 + r^2/k^2)}$$ (II.26)

Furthermore, when sliding halts prior to separation, no changes occur to either the tangential velocity or the angular velocity from this point until the point of separation, that is, $(v_{fT} = v_{sT})$ and $(\omega_f = \omega_s)$.

Expressions for the final tangential and normal velocity components at the centre of mass are:

$$V_{fT} = V_{iT} + \mu\frac{p_{sN}}{p_{cN}}\,V_{iN}$$ (II.27)

$$= V_{iT} - \frac{(V_{iT} + r \cdot \omega_i)}{(1 + r^2/k^2)}$$

$$V_{fN} = -V_{iN} \cdot e_N$$ (II.28)

and the final rotational velocity is:

$$r \cdot \omega_f = r \cdot \omega_i + \mu\frac{r^2}{k^2}\frac{p_{sN}}{p_{cN}}\,V_{iN}$$ (II.29)

$$\omega_f = \omega_i - \frac{r}{k^2}\frac{(V_{iT} + r \cdot \omega_i)}{(1 + r^2/k^2)}$$

Equations (II.27) and (II.28) for the final tangential and normal velocity components, respectively, can then be solved to find the final restitution velocity V_f and angle θ_f as follows:

$$V_f = \sqrt{V_{fT}^2 + V_{fN}^2}$$ (II.30)

$$\theta_f = a\tan\left(\frac{V_{fN}}{V_{fT}}\right)$$ (II.31)

Figure II.4 shows these equations for the final velocities and angles diagrammatically, in terms of the three impact parameters: V_i, θ_i, and ω_i.

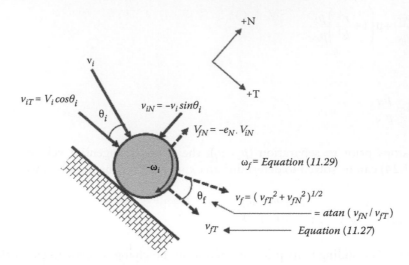

Figure II.4 Diagram of impact showing impact and restitution velocity vectors at the centre of mass and equations for calculating final velocities.

Appendix III: Energy Loss Equations

This appendix shows the derivation of equations defining the energy changes that occur during normal impact of a nonrotating body, and is an expansion of the equations described in Section 6.1 in the main text.

Equations for the energy changes during impact can be developed from the $[p_N - v]$ plot illustrated in Figure III.1 where the impact process is divided into compression and restitution phases. The energy change during each phase is represented by the areas on the $[\delta - F]$ plot or the $[p_N - v]$ plot. The appendix derives the energy changes in terms of the normal impulse p_N and relative velocity v using the linear relationship between these two parameters established in Appendix I, Equation (I.5).

The impact process can be simulated by an infinitesimal deformable particle at the impact point. During impact, the energy E_N generated in the particle by the normal component of the force F_N can be calculated from the relationship between the force and the differen-

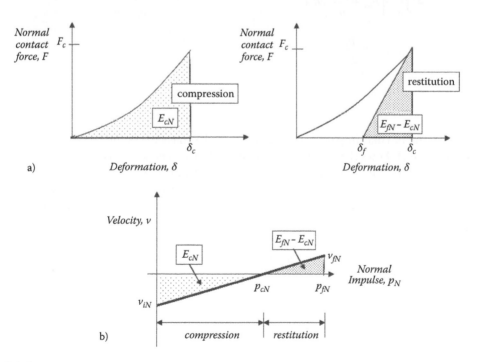

Figure III.1 Energy changes (normal) during compression and restitution phases of impact. (a) Forces generated at contact point during normal impact, energy changes plotted on [force, F – deformation, δ] graph, (b) energy changes plotted on [normal impulse, p_N – velocity, v] graph.

tial normal impulse: $dp_N = F_N \, dt = m \, dv$, so that the energy generated from the moment of impact ($t = i$) up to time t is:

$$E_N = \int_i^t F_N \, v_N \, dt = \int_i^{pN} v_N \, dp_N \qquad \text{(III.1)}$$

For the compression phase of the impact up to the impulse p_{cN}, the relationships between impulse and velocity are as follows:

$$p_{cN} = -m \cdot v_{iN} \qquad \text{(see Equation [I.6])}$$

where v_{iN} is negative because it acts toward the slope in the direction of the ($-N$) axis (see Figure II.1a), and

$$v_N = \left(v_{iN} + \frac{p_N}{m} \right) \qquad \text{(see Equation [I.5])}$$

Therefore, the energy lost during the compression phase of the impact is given by

$$\qquad \qquad \qquad \qquad \qquad \qquad \qquad \qquad \qquad \text{(III.2)}$$

$$
\begin{aligned}
E_N(p_{cN}) &= \int_i^{p_{cN}} v_N \, dp_N \\
&= \int_0^{p_{cN}} \left(v_{iN} + \frac{p_N}{m} \right) dp_N \\
&= v_{iN} \cdot p_{cN} + \frac{p_{cN}^2}{2m} \\
&= -\frac{1}{2} m \cdot v_{iN}^2
\end{aligned}
$$

where v_{iN} is the normal impact velocity, and m is the mass of the body.

Equation (III.2) shows that all the normal impact kinetic energy is lost ($E_N(p_{cN})$ is negative) up to the point of maximum compression, δ_c when the normal velocity is reduced to zero ($v_N = 0$).

A similar approach can be used to find the energy recovered during the restitution phase of the impact, ($E_N(p_f) - E_N(p_c)$) between maximum compression (p_{cN}) and the end of the impact (p_{fN}):

$$
\begin{aligned}
E_N(p_{fN}) - E_N(p_{cN}) &= \int_{p_{cN}}^{p_{fN}} \left(v_{iN} + \frac{p_N}{m} \right) dp_N \\
&= \left[v_{iN} \cdot p_{fN} + \frac{p_{fN}^2}{2m} \right] - \left[v_{iN} \cdot p_{cN} + \frac{p_{cN}^2}{2m} \right] \\
&= \frac{m \cdot v_{iN}^2}{2} \left(1 - \frac{p_{fN}}{p_{cN}} \right)^2
\end{aligned}
$$

The energy recovered during restitution is termed the elastic strain energy. A complete derivation of Equation (III.3) is given in Appendix I, Equation (I.9).

Alternatively, the elastic strain energy can be calculated from the area of the restitution triangle between impulse values p_{cN} and p_{fN} on Figure III.1b as follows:

$$E_N(p_{fN}) - E_N(p_{cN}) = \frac{1}{2}(p_{fN} - p_{cN})\, v_{fN} \tag{III.3}$$

$$= \frac{1}{2}\, p_{cN}\left(\frac{p_{fN}}{p_{cN}} - 1\right)\left(v_{iN} + \frac{p_{fN}}{m}\right)$$

$$= -\frac{m \cdot v_{iN}}{2}\left(\frac{p_{fN}}{p_{cN}} - 1\right)\left(v_{iN} - \frac{p_{fN} \cdot v_{iN}}{p_{cN}}\right)$$

$$= \frac{m \cdot v_{iN}^2}{2}\left(1 - \frac{p_{fN}}{p_{cN}}\right)\left(1 - \frac{p_{fN}}{p_{cN}}\right)$$

$$= \frac{m \cdot v_{iN}^2}{2}\left(1 - \frac{p_{fN}}{p_{cN}}\right)^2$$

Equations (II.2) and (III.3) together define the net energy loss during normal impacts as;

$$E_N(net) = [\text{energy lost in compression}] + [\text{energy gained in restitution}] \tag{III.4}$$

$$= \left[E_N(p_{cN})\right] + \left[E_N(p_{fN}) - E_N(p_{cN})\right]$$

$$= -\frac{m \cdot v_{iN}^2}{2} + \frac{m \cdot v_{iN}^2}{2}\left(1 - \frac{p_{fN}}{p_{cN}}\right)^2$$

$$= -\frac{m \cdot v_{iN}^2}{2}\left[1 - \left(1 - \frac{p_{fN}}{p_{cN}}\right)^2\right]$$

Equation (I.12) defines the relationship between normal impulses p_{fN}, p_{cN}, and the normal coefficient of restitution e_N as:

$$p_{fN} = -m \cdot v_{iN}(1 + e_N) = p_{cN}(1 + e_N) \tag{III.5}$$

and

$$\frac{p_{fN}}{p_{cN}} = (1 + e_N)$$

and

$$e_N^2 = \left(1 - \frac{p_{fN}}{p_{cN}}\right)^2$$

Substitution of Equation of (III.5) into Equation (III.4) gives the following expression for the net energy loss during normal impact:

$$E_N(net) = -\frac{1}{2}m \cdot v_{iN}^2 \ (1 - e_N^2)$$

(III.6)

In the development of Equation (III.6) for normal impact, the value of e_N is always less than 1 because it is defined by the energy losses for a spherical body.

Appendix IV: Conversion Factors

Imperial Unit	SI Unit	SI Unit Symbol	Conversion Factor(Imperial to SI)	Conversion Factor(SI to Imperial)
Length				
Mile	kilometer	km	1 mile = 1.609 km	1 km = 0.6214 mile
Foot	meter	m	1 ft = 0.3048 m	1 m = 3.2808 ft
	millimeter	mm	1 ft = 304.80 mm	1 mm = 0.003281 ft
Inch	millimeter	mm	1 in = 25.40 mm	1 mm = 0.03937 in
Area				
Square mile	square kilometer	km^2	1 $mile^2$ = 2.590 km^2	1 km^2 = 0.3861 $mile^2$
	hectare	ha	1 $mile^2$ = 259.0 ha	1 ha = 0.003 861 $mile^2$
Acre	hectare	ha	1 acre = 0.4047 ha	1 ha = 2.4710 acre
	square meter	m^2	1 acre = 4,047 m^2	1 m^2 = 0.000247 1 acre
Square foot	square meter	m^2	1 ft^2 = 0.09290 m^2	1 m^2 = 10.7639 ft^2
Square inch	square millimeter	mm^2	1 in^2 = 645.2 mm^2	1 mm^2 = 0.001550 in^2
Volume				
Cubic yard	cubic meter	m^3	1 yd^3 = 0.7646 m^3	1 m^3 = 1.3080 yd^3
Cubic foot	cubic meter	m^3	1 ft^3 = 0.02832 m^3	1 m^3 = 35.3147 ft^3
	liter	l	1 ft^3 = 28.32 l	1 liter = 0.03531 ft^3
Cubic inch	cubic millimeter	mm^3	1 in^3 = 16 387 mm^3	1 mm^3 = 61.024 $\times$ 10^{-6} in^3
	cubic centimeter	cm^3	1 in^3 = 16.387 cm^3	1 cm^3 = 0.06102 in^3
	liter	l	1 in^3 = 0.01639 l	1 liter = 61.02 in^3
Imperial gallon	cubic meter	m^3	1 gal = 0.00455 m^3	1 m^3 = 220.0 gal
	liter	l	1 gal = 4.546 l	1 liter = 0.220 gal
Pint	liter	l	1 pt = 0.568 l	1 liter = 1.7598 pt
US gallon	cubic meter	m^3	1 US gal = 0.0038 m^3	1 m^3 = 264.2 US gal
	liter	l	1 US gal = 3.8 l	1 liter = 0.264 US gal
Mass				
Ton	tonne	t	1 ton = 0.9072 tonne	1 tonne = 1.1023 ton
Ton (2,000 lb) (US)	kilogram	kg	1 ton = 907.19 kg	1 kg = 0.001102 ton
Ton (2,240 lb) (UK)	kilogram	kg	1 ton = 1,016.0 kg	1 kg = 0.000984 ton
Kip	kilogram	kg	1 kip = 453.59 kg	1 kg = 0.0022046 kip
Pound	kilogram	kg	1 lb = 0.4536 kg	1 kg = 2.2046 lb

(continued)

Imperial Unit	SI Unit	SI Unit Symbol	Conversion Factor (Imperial to SI)	Conversion Factor (SI to Imperial)
Mass Density				
ton per cubic yard (2,000 lb) (US)	kilogram per cubic meter	kg/m³	1 ton/yd³ = 1186.55 kg/m³	1 kg/m³ = 0.0008428 ton/yd³
	tonne per cubic meter	t/m³	1 ton/yd³ = 1.1866 t/m³	1 t/m³ = 0.8428 ton/yd³
ton per cubic yard (2,240 lb) (UK)	kilogram per cubic meter	kg/cm³	1 ton/yd³ = 1328.9 kg/m³	1 kg/m³ = 0.00075 ton/yd³
pound per cubic foot		t/m³	1 lb/ft³ = 16.02 kg/m³	1 kg/cm³ = 0.06242 lb/ft³
	tonne per cubic meter		1 lb/ft³ = 0.01602 t/m³	1 t/m³ = 62.42 lb/ft³
pound per cubic inch	gram per cubic centimeter	g/cm³	1 lb/in³ = 27.68 g/cm³	1 g/cm³ = 0.03613 lb/in³
	tonne per cubic meter	t/m³	1 lb/in³ = 27.68 t/m³	1 t/m³ = 0.03613 lb/in³
Force				
ton force (2,000 lb) (US)	kilonewton	kN	1 tonf = 8.896 kN	1 kN = 0.1124 tonf (US)
ton force (2,240 lb) (UK)	kilonewton	kN	1 tonf = 9.964 KN	1 kN = 0.1004 tonf (UK)
kip force	kilonewton	kN	1 kipf = 4.448 kN	1 kN = 0.2248 kipf
pound force	newton	N	1 lbf = 4.448 N	1 N = 0.2248 lbf
tonf/ft (2,000 lb) (US)	kilonewton per meter	kN/m	1 tonf/ft = 29.189 kN/m	1 kN/m = 0.03426 tonf/ft (US)
tonf/ft (2,240 lb) (UK)	kilonewton per meter		1 tonf/ft = 32.68 kN/m	1 kN/m = 0.0306 tonf/ft (UK)
pound force per foot	newton per meter	N/m	1 lbf/ft = 14.59 N/m	1 N/m = 0.06852 lbf/ft
Flow Rate				
cubic foot per minute	cubic meter per second	m³/s	1 ft³/min = 0.000 471 9 m³/s	1 m³/s = 2118.880 ft³/min
	liter per second	l/s	1 ft³/min = 0.4719 l/s	1 l/s = 2.1189 ft³/min
cubic foot per second	cubic meter per second	m³/s	1 ft³/s = 0.028 32 m³/s	1 m³/s = 35.315 ft³/s
	liter per second	l/s	1 ft³/s = 28.32 l/s	1 l/s = 0.03531 ft³/s
gallon per minute	liter per second	l/s	1 gal/min = 0.075 77 l/s	1 l/s = 13.2 gal/min
Pressure, Stress				
ton force per square foot (2,000 lb) (US)	kilopascal	kPa	1 tonf/ft² = 95.76 kPa	1 kPa = 0.01044 ton f/ft²
ton force per square foot (2,240 lb) (UK)	kilopascal	kPa	1 tonf/ft² = 107.3 kPa	1 kPa = 0.00932 ton/ft²

(continued)

Imperial Unit	SI Unit	SI Unit Symbol	Conversion Factor(Imperial to SI)	Conversion Factor(SI to Imperial)
pound force per square foot	pascal	Pa	1 lbf/ft^2 = 47.88 Pa	1 Pa = 0.020 89 lbf/ft^2
	kilopascal	kPa	1 lbf/ft^2 = 0.047 88 kPa	1 kPa = 20.89 lbf/ft^2
pound force per square inch	pascal	Pa	1 lbf/in^2 = 6895 Pa	1 Pa = 0.0001450 lbf/in^2
	kilopascal	kPa	1 lbf/in^2 = 6.895 kPa	1 kPa = 0.1450 lbf/in^2
Weight Density[a]				
pound force per cubic foot	kilonewton per cubic meter	kN/m^3	1 lbf/ft^3 = 0.157 kN/m^3	1 kN/m^3 = 6.37 lbf/ft^3
Energy				
Foot lbf	joules	J	1 ft lbf = 1.356 J	1 J = 0.7376 ft lbf

[a] Assuming a gravitational acceleration of 9.807 m/s^2.

References

Ashayer, P. (2007). Application of Rigid Body Impact Mechanics and Discrete Element Modeling to Rockfall Simulation. Ph.D. thesis, University of Toronto, Grad. Dept. of Civil Engineering.

Asteriou, H., Saroglou, G., Tsiambaus, T. (2012). Geotechnical and kinematic parameters affecting the coefficient of resolution for rockfall analysis. *Int. J. Rock Mech. Min. Sci.* 54, 103–113.

Beladjine, D., Amimi, M., Oger, L., and Valance, A. (2007). Collision process between an incident bead and a three-dimensional grandual packing. *Physiol. Rev. E*, 75, 1–12.

Benjamin, J. R. and Cornell, A. C. (1979). *Probability, Statistics and Decision for Civil Engineers.* McGraw-Hill, New York.

Bichler, A. and Stelzer, G. (2012). A fresh approach to the hybrid/attenuator rockfall fence. *Landlsides and Engineered Slopes.* E. Eberhardt, C. Froese, K. Turner, and S. Leroueil, CRC Press. June 2–8, Banff, Canada.

Brunet, G. and Giacchetti, G. (2012). Design software for secured drapery. *Proc. 63rd Highway Geology Symp.*, Redding, CA.

Bugnion, L., Wendeler, C., and Shevlin, T. (2012). Impact pressure measurements in shallow landslides for flexible barrier design. *Proc. 11th International Symposium on Landslides (ISL) and the 2nd North American Symposium on Landslides,* June 2–8, 2012, Banff, Alberta, Canada.

Bunce, C. M., Cruden, D. M., and Morgenstern, N. R. (1997). Assessment of the hazard from rock fall on a highway. *Canadian Geotech. J.*, 34, 344–356.

Buzzi, O., Giacomini, A., and Spadari, M. (2012). Laboratory investigation of high values of restitution coefficients. *J. Rock Mech. Rock Eng.*, 43, 35–43.

California Polytechnical State University (1996). Response of the Geobrugg Cable Net System to Debris Flow Loading. Department of Civil and Environmental Engineering, San Luis Obispo, CA, pp. 3, 35, 59.

Canadian Geotechnical Society (2006). *Canadian Foundation Engineering Design Manual*, 4th edition. BiTech Publishing, Richmond, British Columbia, 488 pages.

Chau, K. T., Wong, R. H. C., Liu, J., and Lee, C. F. (2003). Rockfall hazard analysis for Hong Kong based on rockfall inventory. *Rock Mech Rock Eng*, 36(5), 383–408.

Chau, K. T., Wong, R. H. C., and Wu, J. J. (2002). Coefficient of restitution and rotational motions of rock fall impacts. *Int. J. Rock Mech. Mining Sci*, 39, 69–77.

Chen, G., Zheng, L., Zhang, Y., and Wu, J. (2013). Numerical simulation in rockfall analysis: A close comparison of 2-D and 3-D DDA. *Rock Mech. Rock Eng.*, 46(3), 527–541.

Dellow, G., Yetton, M., Archibald, G., Barrell, D. J. A., Bell, D., Bruce, Z., Campbell, A., Davies, T., De Pascale, G., Easton, M., Forsyth, P. J., Gibbons, C., Glassey, P., Grant, H., Green, R., Hancox, G., Jongens, R., Kingsbury, P., Kupec, J., MacFarlane, D., McDowell, B., McKelvey, B., McCahon, I., McPherson, I., Molloy, J., Muirson, J., O'Halloran, M., Perrin, N., Price, C., Read, S., Traylen, N., Van Dissen, R., Villeneuve, M., and Walsh, I. (2011) Landslides caused by the 22 February 2011 Christchurch earthquake and management of landslide risk in the immediate aftermath. *Bull. New Zealand Soc. Earthquake Eng.*, 44(4), 227–238.

Dijkstra, T. A., Chandler, J., Wackrow, R., Meng, X. M., Ma, D. T., Gibson, A., Whitworth, M., Foster, C., Lee, K., Hobbs, P. R. N., Reeves, H. J., and Wasowski, J. (2012). Geomorphic controls and debris flows—the 2010 Zhouqu disaster, China. *Proc. 11th Int. Symp. Landslides (ISL) and the 2nd North American Symposium on Landslides,* June 2–8, 2012, Banff, Alberta, Canada.

Dorren, L. K. A. and Berger, F. (2012). Integrating forests in the analysis and management of rock fall risks: experience from research and practice in the Alps. *Proc. 11th Int. Conf. and 2nd North American Symp. on Landslides and Engineered Slopes.* Banff, Canada, June 2012, pp. 117–127.

Dorren, L. K. A., Berger, F., and Putters, U. S. (2006). Real size experiments and 3-D simulation of rock fall on forested and non-forested slopes. *Natural Hazards Earth Sci. Syst,* 6, 145–153.

Dorren, L. K. A. and Berger, F. (2005). Stem breakage of trees and energy dissipation during rock fall events. *Tree Physiol,* 26, 63–71.

European Organisation for Technical Approvals (February 1, 2008). ETAG 27, Guideline for European Technical Approval of Falling Rock Protection Kits. EOTA. Retrieved on May 22, 2013, from http://www.eota.be/pages/home/.

Geobrugg (2012). RUVOLUM Online Tool, Version 2012 software. Geobrugg AG, Switzerland. Retrieved on May 22, 2013, from http://applications.geobrugg.com/

Geobrugg (2010). Product Literature—TECCO Damping Modules for Rock Fall Protection Galleries. Romanshorn, Switzerland.

Giacchetti, G., Grimod, A., and Cheer, D. (2011). Soil nailing with flexible facing: design and experience. *Proc. Second World Landslide Forum,* Rome, October.

Giacchetti, G. and Zotti. I. M. (2012). Design approach for rockfall barriers. *Geotechnical XI National Congress in Costa Rico* (August 9 and 10). San Jose, Costa Rica, 2012.

Giacomini, A., Thoeni, K., Lambert, C., Booth, S., and Sloan, S. W. (2012). Experimental study on rockfall drapery systems for open pit highwalls. *Int. J. Rock Mech. Mining Sci.* 56(0), 171–181.

Giacomini, A., Spadari, M., Buzzi, O., Fityus, S. G., and Giani, G. P. (2010). Rockfall motion characteristics on natural slopes in eastern Australia. Euroch 2010, Lausazine, Switzerland.

Goldsmith, W. (1960). *Impact—Theory and Physical Behaviour of Colliding Solids* (2001 edition). Dover Publications, New York, 379 pages.

Grimod, A. and Giacchetti, G. (2011). Protection from high energy impacts using reinforced earth embankments: Design and experience. *Proc. Second World Landslide Forum,* Rome, October.

Gutenberg, B. and Richter, C. F. (1954). *Seismicity of the Earth,* 2nd ed., Princeton University Press, Princeton, NJ.

Hambleton, J. P., Buzzi, O., Giacomini, A., Spadari, M., and Sloan, S. (2013). Perforation of flexible rockfall barriers by normal block impact. *Rock Mech. Rock Eng.,* 46(3), 515–526.

Harp, E. L., Jibson, R. W., Kayen, R. E., Keefer, D. K., Sherrod, B. L., Carver, G. A., Collins, B. D., Moss, R. E. S., and Sitar, N. (2003). Landslides and liquefaction triggered by the M7.9 Denali Fault earthquake of 3 November, 2002. *GSA Today,* August 2003, 10 pages.

Harp, E. L. and Jibson, R. W. (1995). Inventory of Landslides Triggered by the 1994 Northridge, California Earthquake. Dept. of the Interior, USGS, Open file Report 95-213, 17 pages.

Harp, E. L. and Wilson, R. C. (1995). Shaking intensity thresholds for rock falls and slides: Evidence from 1987 Whittier Narrows and Superstition Hills earthquakes strong motion records. *Bull. Seismological Soc. Am.,* 85(6), 1739–1757.

Harr, M. E. (1977). *Mechanics of Particulate Matter—A Probabilistic* Approach. McGraw-Hill, New York, 543 pages.

Hungr, O. and Evans, S. G. (1988). Engineering evaluation of fragmental rock fall hazards. *Proc. 5th Inter. Symp. on Landslides,* Lausanne, Switzerland, July, 685–90.

Hungr, O., Evans, S. G., and Hazzard, J. (1999). Magnitude and frequency of rock falls and rock slides along the main transportation corridors on southwestern British Columbia. *Can. Geotech. J.,* 36, 224–238.

Jacquemoud, J. (1999). Swiss Guideline for the design of rock fall protection galleries: background, safety concept and case histories. *Proc. Joint Japan-Swiss Seminar on Impact Load by RockFalls and Design of Protection Measures.* Kanazawa, Japan, October, 161 pages.

Japan Road Association (2000). *Rock Fall Control Manual* (in Japanese), 422 pages.

Jibson, R. W. and Harp, E. L. (1995). Inventory of Landslides Triggered by the 1984 Northridge, California Earthquake. U.S. Dept. of the Interior, USGS, Open file 95-213, 17 pp.

Jones, C. L., Higgins, J. D., and Andrew, R. D. (2000). Colorado Rockfall Simulation Program, version 4.0 (CRSP). CO Dept. of Transportation, Denver, CO, Report No. CDOT-STMB-CGS-99-1, 127 pages.

Keefer, D. K. (1992). The susceptibility of rock slopes to earthquake-induced failure. *Proc. 35th Annual Meeting of the Assoc. of Eng. Geologists* (ed. Martin L. Stout), Long Beach, CA, pp. 529–538.

Kirsten, H. A. D. 1982. Design and construction of the Konwyn's Pass rock fall shelter. *Civ. Engr. S. Afr*, 24(9), 477–492.

Kobayashi, Y., Harp, E. L., and Kagawa, T. (1990). Simulation of rock falls triggered by earthquakes. *J. Rock Mech. Rock Eng.*, 23(1), 1–20.

Komura, T., Muranishi, T., Nisizawa, K., and Masuya, H. (2001). Study on parameters concerning impact of falling rock on field slopes and rock fall simulation method. *Proc. 4th Asia-Pacific Conf. on Shock and Impact Loads on Structures*, 345–352.

Labiouse, V., Descoeudres, F., and Montani, S. (1996). Experimental study of rock sheds imapcted by rock blocks. *Struct. Eng. Int.*, 3, 171–176.

Maccaferri S. p. A, Bologna, Italy (2012). Rockfall and Debris Flow Protection Embankments. Product literature, www.maccaferri.ca.

Maccaferri S. p. A. (2012). *Slope mesh design software, BIOS and Mac.Ro1*. Maccaferri S. p. A., Bologna, Italy.

Maggs, M. (2007, October 8). Bouncing Ball Stobe Edit, in Wikimedia Commons. Retrieved May 22, 2013, from https://en.wikipedia.org/wiki/File:Bouncing_ball_strobe_edit.jpg

Mamaghani, I. H. P., Yoshida, H., and Obata, Y (1999). Reinforced expanded polystyrene Styrofoam covering rock sheds under impact of falling rocks. Joint Japan-Swiss Scientific seminar on impact loads by rock falls and design of protective structures, Kanazawa, Japan, October, 79–89.

Masuya, H., Amanuma, K., Nishikawa, Y., and Tsuji, T. (2009). Basic rock fall simulation with consideration of vegetation and application of protection measures. *Nat. Hazards and Earth Systems Sciences*, 9, 1835–1843.

Masuya, H., Ihara, T., Onda, S., and Kamijo, A. (2001). Experimental study on some parameters for simulation of rock fall on slope. *Proc. 4th Asia-Pacific Conf. on Shock and Impact Loads on Structures*, 63–69.

Masuya, H., Maegawa K., Mizuki A., and Yoshida, H. 1987. Impulsive loads by rock falls on steel rock-sheds (in Japanese), *J. Struct. Eng.*, 36A, 41–49, March.

McCauley, M. L., Works, B. W., and Naramore, S. A. (1985). Rock Fall Mitigation. Report FHWA/CA/TL-85/12. FHWA, U.S. Department of Transportation.

National Cooperative Highway Research Program (NCHRP). (1999). Geotechnical related developments of load and resistance factor design (LRFD) synthesis of highway practice (276), TRB, Washington, D.C., 69 p.

Newmark, N. M. (1965). Effects of earthquakes on dams and enbankments. *Geotechnique*, 15(2), 139–160.

Newton, I. (1687). *Philosophiae Naturalis Principia Mathematica*. Cambridge University Press, Cambridge.

Nishikawa, Y., Masuya, H., and Moriguti, Y. (2012). Three dimensional simulation of rock fall motion with consideration of roughness of the slope surface. *Trans. Japan Soc. for Computational Engineering and Science*, 2012, 20120003.

Nocilla, N., Evangelista, A., and di Santolo, A. S. (2009). Fragmentation of rock falls: Two Italian case studies of hard and soft rocks. *Rock Mech. Rock Eng.*, 42, 815–833.

Palisades Corporation (2012). Computer program @RISK, version 5.0. Palisade Corporation, Ithaca, New York.

Patton, F. D. (1974). Multiple modes of shear failure in rock. *Proc. 1st. Int. Cong. on Rock Mechanics*, Lisbon, 1, 509–513.

Peckover, F. L. (1975). Treatment of rock falls on railway lines. *American Railway Engineering Association*, Bulletin 653, Washington, DC, pp. 471–503.

Pfeiffer, T. J. and Bowen, T. D. (1989). Computer simulation of rock falls. *Bull. Assoc. Engineering Geologists*, 36(1), 135–146.

Pfeiffer, T. J., Higgins, J. D., Andrew, R. D., Schultz, R. J. and Beck, R. B. (1995). Colorado Rock Fall Simulation Program. Version 3.0, User's Manual. Colorado Department of Transportation.

Pfeiffer, T. J. and Higgins, J. D. (1995). *Rockfall Hazard Analysis Using the Colorado Rockfall Simulation Program*. TRB, Washington, DC, Transportation Research Record 1288, 117–126.

Pierson, L. A., Davis, S. A., and Van Vickle, R. (1990). *The Rockfall Hazard System, Implementation Manual*. Technical Report, #FHWA-OR-EG-90-01, Washington, DC.

Pierson, L. A., Gullixson, C. F., and Chassie, R. G. (2001). *Rockfall Catchment Area Design Guide*. Research Report SPR-3(032): Oregon Department of Transportation-Research Group, Federal Highway Administration.

Piteau and Associates Ltd. (1980). Slope stability analysis for rockfall problems: The computer rockfall model for simulating rockfall distributions, Part D. *Rock Slope Engineering*, reference manual FHWA-TS-79-208. FHWA, Department of Transportation, Washington, D.C., 62–68.

Poisson, S. D. (1811). *Traité de Méchanique*. Courier, Paris (English translation by Rev. H. H. Harte, Longman & Co 1817).

Protec Engineering, Niigata, Japan (2012). MSE Embankments for Rock Fall Containment. Technical literature, www.proteng.co.jp.

Raiffa, H. (1968). *Decision Analysis: Introductory Lectures on Choices under Uncertainty*. Addison-Wesley, Reading, MA, 309 pages.

Ritchie, A. M. (1963). An evaluation of rock fall and its control. *Highway Research Record* 17, Highway Research Board, NRC, Washington DC, pp. 13–28.

RGHRG (Railway Ground hazard Research Group), 2012. Annual Meeting, Vancouver, Canada, December.

Rochet, L. (1987). Application des modeles numeriques de propagation a l'tude des eboulements rocheux. *Bulletin de liaison laboratorires des ponts et chaussees*. 150–151, Sept.–Oct.

RocScience (2012). Computer program *RocFall*. RocScience Inc., Toronto, Canada.

Schellenberg, K. and Vogel, T. (2009). A dynamic design method for rockfall protection galleries. *J. Int. Assoc. Bridge and Structural Eng.*, SEI 19(3), 321–326.

Skermer, N. A. (1984). M Creek debris flow disaster. *Canadian Geotechnical Conference: Canadian Case Histories, Landslides*, Toronto, pp. 187–194.

Spadari, M., Giacomini, A., Buzzi, O., Fityus, S. G., and Giani, G. P. (2012). In situ rockfall testing in New South Wales, Australia. *Int. J. Rock Mech. Min. Sci.* 49, 84–93.

Stevens, W. D. (1998). RocFall: A Tool for Probabilistic Analysis, Design of Remedial Measures and Prediction of Rock Falls. Thesis for M. App. Sc. degree, University of Toronto, Dept. of Civil Eng., 28 pages.

Stock, G. M., Luco, N., Harp, E. L., Collins, B. D., Reichenbach, P., Frankel, K. L., Matasci, B., Carrea, D., Jaboyedoff, M., and Oppikofer, T. (2012). Quantitative rock fall hazard and risk assessment in Yosemite Valley, California, USA. *Landslides and Engineered Slopes: Protecting Society through Improved Understanding*, 2, 1119–1125.

Stock, G. M., Sitar, N., Borchers, J. W., Harp, E. L., Snyder, J. B., Collins, B. D., Bales, R. C., and Wiezorek, G. F. (2012). Evaluation of hypothesized water-system triggers for rock falls from Glacier Point, Yosemite National Park, California, USA. *Landslides and Engineered Slopes: Protecting Society through Improved Understanding*, 2, 1165–1171.

Stronge, W. J. (2000). *Impact Mechanics*. Cambridge University Press, Cambridge, U.K., 280 pages.

Transportation Research Board (TRB) (1996). *Landslides, Investigation and Mitigation*. National Research Council, Special Report 247, Washington, D.C., 673 pages.

Ushiro, T. and Tsutsui, H. (2001). Movement of a Rockfall and a Study on Its Prediction. Roc. Int. Symp. Geotech. Environ. Challenges Mount. Terrain, Ehime, Japan, Nov. 2001, 275–285.

Ushiro, T., Kusumoto, M., Shinohara, S., and Kinoshita, K. (2006). An experimental study related to rock fall movement mechanism. *J. Japan Soc. Civil Engineers*, Series F, 62(2), 377–386 (in Symp. on Geotechnical and Environmental Challenges in Mountainous Terrrain, Kathmandu, Nepal, 366–375.

Ushiro, T., Ohara, K., Akisaka, N., and Yoshizaki, K. (1983). The study of the motion of falling rock using the experimental data (in Japanese). Report of Second Symp. on Impact due to RockFalls and Design of Rock Sheds, July, 23–35

Vogel, T., Labiouse, V., and Masuya, H. (2009). Rockfall protection as an integral task. *Structural Engineering International*, SEI 19(3), 304–312, IABSE, Zurich, Switzerland, www.iabse.org

Watts, C. F., Underwood, S. A., Haneberg, W. C., and Rogers, J. D. (2012). Fully rationalized equations for incorporating joint water pressure in rock slope stability analyses at Glacier Point in Yosemite National Park, California. *Landslides and Engineered Slopes: Protecting Society through Improved Understanding*, 2, 1173–1178.

Wu, Shie-Shin (1985). Rockfall Evaluation by Computer Simulation, Transportation Research Record. 1031, 1–5.

Wyllie, D. C. (1987). Rock slope inventory system. Proc. Federal Highway Administration Rock Fall Mitigation Seminar, FHWA, Region 10, Portland, OR.

Wyllie, D. C. (1999). *Foundations on Rock* (2nd edition). E&FN Spon, London, 401 pages.

Wyllie, D. C. and Mah, C. W. (2002). *Rock Slope Engineering*, 4th edition. Taylor & Francis, London, 422 pages.

Wyllie, D. C. (2006). Risk management of rock fall hazards. *Proc. 59th Annual Canadian Geotechnical Conference*, Canadian Geotech. Soc., Vancouver, Canada.

Wyllie, D. C., McCammon, N. R., and Brumund, W. F. (1979). *Use of Risk Analysis in Planning Slope Stabilization Programs on Transportation Routes*. Transportation Research Board, Research Record 749, Washington, D.C.

Xcitex (2008). Proanalyst©, Motion Analysis Software. Xcitex Inc., Cambridge, MA.

Yokoyama, K., Ohira, M., and Yoshida, H. (1993). Impulsive properties due to rock fall on a steel rock shed (in Japanese), *J. Struct. Eng.*, 39A, 1573–1586.

Yoshida, H., Matuba, Y., Hohki, K., and Kubota T. (1991a). An experimental study on shock absorbing effect of expanded polystyrene against falling rocks (in Japanese), *Proc. Japan Society of Civil Eng.*, 427, 143–152.

Yoshida, H. Masuya, H., Satou, M., and Ihara, T. (1987). A database of rock falling tests and an estimation of impulsive load by rock falls (in Japanese), *J. Struct. Eng.*, 33A.

Yoshida, H., Ushiro, T., Masuya, H., and Fujii, T. (1991b). An evaluation of impulsive design load of rock sheds taking into account slope properties (in Japanese), *J. Struct. Eng.*, 37A, 1603–1616.

Yoshida, H., Nomura, T., Wyllie, D. C., and Morris, A. J. (2007). Rock fall sheds—application of Japanese designs in North America. *Proc. 1st. North American Landslide Conference*, Vail, CO., AEG Special Publication No. 22., ed. Turner, A. K., Schuster, R. L., pp. 179–196.

Zhang, R. and Rock, A. (2012). Rockfall stimulation using a 3-D discrete element model with engineering validation. *Landslides and Engineered Slopes: Protecting Society through Improved Understanding*, 2, 1153–1164.

Index

Printed and bound by CPI Group (UK) Ltd, Croydon, CR0 4YY

18/10/2024

01776249-0007